1

Going to College?

Questions
Every Christian Should Be Able To Answer

J Philip A Hinton, MD

phintonmd@gmail.com

Third Edition January 2018

Cover design, art and illustrations by
Julie Elizabeth Myers Hinton 2018
jemyers8@gmail.com

CONTENTS

Introduction

This is a college preparation textbook for 11th and 12th graders. The reason for this book is that serious questions about your faith arise when you leave home and begin living independently. Many young Christians attended church because their families did. They were raised in church and attended youth groups. They went to Christian camps. But their faith was not their own. It was adopted from their parents. It was part of their home life and culture. And when challenged by exposure to opposing views on college campuses they had no viable answers to why they believe what they believe. These questions need to be asked and answered before they leave for college. These young Christians need to decide for themselves if Christian beliefs are true. And if they decide they are true they need to put their faith into action.

The Big 3 Doubt Drivers
- ## Is God real?
- ## Is there reasonable evidence for God?
- ## How could a good God allow suffering and evil?

The first question they encounter is, "Is God real? Doesn't science prove there is no God? Doesn't science show that the Big Bang and Evolution can explain the origin of life?" Either Jesus and God are real—factual—or not. This is of the utmost importance. There is no benefit in believing a lie, either believing in God if there is no God, or believing in atheism if there really is a God. If God isn't real why shouldn't they leave their faith? And if God is real they have to decide if they are willing to submit to His authority. It's the most important decision they will ever make.

Answering the question "Is God real?" is the purpose of Chapters 1 and 2. The notion that science can "prove" there is no god is simply wrong. Science can neither prove nor disprove the existence of God. Both belief in God and belief in naturalism (the worldview that there is no god and that the universe and life developed by chance) are not falsifiable. This means that neither view can be proven wrong. Scientific theories or hypotheses are guesses about origins based on the best available evidence. The hypothesis about how the universe came into being is called the Big Bang *Theory*. And it is the *Theory* of Abiogenesis and Evolution. As more data become available scientific theories are either confirmed as the best explanation for the findings or they are dropped for better explanations.

The thesis of this book is simple. There is overwhelming evidence for the reality of God. The weight of scientific evidence strongly supports design rather than chance as the origin of the universe and life. Design requires a designer. This is in total opposition to the dominant beliefs of our culture—that there is only the "appearance of design," and that our universe and life are the result of chance.

There is overwhelming scientific evidence for design rather than chance as the origin of the universe and life. Design requires a designer.

But despite the scientific evidence for design there are still many scientists and philosophers in universities across America who persist in their "no god" belief. These self-named "new atheists" include Daniel Dennett, Richard Dawkins, Carl Sagan, and Michael Martin. Carl Sagan started each episode of his series "Cosmos" on PBS TV with, "The universe is all there is, all there was, and all there ever will be." These influential spokesmen proclaim their beliefs as authoritative. But are they? As Eric Metaxis points out, these are our culture's version of Pharisees. Just as Pharisees in Jesus time refused to recognize the truth about Jesus because He threatened their preconceived beliefs and their status in society, these "new atheists'" beliefs and status are threatened by the most recent scientific evidence indicating that design is real rather than apparent.

Another question that often follows the first is this. "If belief in God is not falsifiable, and belief that there is no god is not falsifiable, why should I believe in God? Isn't it just a toss up?" The answer to this question can be found in Chapter 2. In addition to the physical scientific evidence, there are multiple human characteristics including built-in moral law and free will that indicate that humans are created in the image of God. Additional evidence is found in the experience of God. This chapter summarizes the multiple lines of evidence that support belief in God—that indicate that belief in God is warranted. Warranted means well supported by the evidence, having enough evidence to act upon.

The next question is philosophical. "How could a God who is good, all knowing, and all powerful allow all this suffering and evil to exist? If there is such a God, why doesn't He do something about child sex trafficking, genocide, ethnic and religious wars, birth defects in babies, tidal waves and devastating earthquakes? If He knows about them, and is able to stop them, why doesn't He?" In fact, the answer to this question is fairly simple. You can find it in Chapter 3. But if you don't know the answer, if you haven't studied this question in advance it seems to be a powerful argument against the existence of God.

If you decide, on the basis of the evidence, that God is real or likely, what should you do?

Once you have decided for yourself that God is real, the next questions are "What is God like?" and "What does He want me to do?" And to get these answers you have to go to the evidence He has provided for us: the Bible and the person of Jesus.

Many people ask this. "Why do you think the Bible is true? Isn't it an ancient document filled with myths and miracles, things that aren't really true and couldn't really happen? Isn't it true that the Bible has been copied over and over so many times that it is full of mistakes?" The answer to this question is in Chapter 4. There is excellent evidence for the historical accuracy and truth of the Bible, but if you don't know what it is it won't help you when you encounter this objection.

Next come questions about Jesus. "I believe that Jesus was an influential person and a powerful teacher, but what makes you think He was actually God? Isn't that just insanity—a man claiming to be God? He didn't really claim to be God, did he? Isn't it true that he was just a man, made into a god after his death by followers who wanted to gain status and power by being the leaders of a new religion?" The answer to this question can be found in Chapter 5. The best kind of historical evidence—eyewitness testimony—backs the claims of Jesus. If you don't know the facts about who Jesus is and the evidence that supports it you can not defend your faith.

You will certainly encounter many who believe that morals are relative. Their question goes something like this. "What makes you so sure that your outdated morals are important? If what I do doesn't hurt someone how can it be wrong?" The answer to this question can be found in Chapter 6. The answer is a question you can ask moral relativists that will upend their belief that morals are relative.

You will run into some who believe that being good is what matters. You may even believe this yourself. "I'm not harming anyone, and I don't think my behavior is any different than anyone else, so why do you think I'm wrong? I don't need forgiveness, so why do I need God? If there really is a god, aren't I as good as anybody else? Don't you think I am good enough to go to heaven?" The answer to this is found in Chapter 7. It's only after you discover for yourself that God's moral law is real and that you have broken it that you understand the danger you are in. If God is real, if the Bible is true, if Jesus represents what God is like, what must you do to be right with God? This is important because if God is the source of life leaving him means you are choosing death. He doesn't want you to choose death. But He gives you free choice. Pascal's wager describes how you bet your life on what you believe, and why what you believe really matters.

Others will say, "The kind of God you describe is not fair. How can it be fair that someone who hasn't heard about God and Jesus goes to hell when they die?" The answer to this can be found in Chapter 8.

Christian beliefs about chastity are unpopular. "Why are you so bound by tradition that you think sex outside marriage is sin? Everybody does it, and it doesn't hurt anybody, so what's wrong with it?" The answer can be found in Chapter 9.

"Abortion is legal in the United States, so why are Christians against it? Why do you think abortion is murder? It's my body, so why can't I do what I like with it? Don't you know you can't legislate morality?" The answer to these questions can be found in Chapter 10.

"Why do you think Christians are better off than people who don't believe in God? Don't Christians suffer and have problems just like everybody else?" This answer can be found in chapter 11.

Finally, lots of educated people ask, "Doesn't the Bible say the earth was created in six days? Why do you believe in a young earth when scientists say it is over 4 billion years old?" The answer to this question is in Chapter 12.

Everyone decides what to believe on the basis of authority. Some of us trust what our teachers or professors tell us is true. Some trust Wikipedia, a textbook, National Geographic magazine, TV news, Facebook, Twitter, or the Bible. You may decide because of what you were told in a natural history museum or planetarium. But you will find it difficult to tell what is fact versus what is just opinion or theory because of the differing truth claims expressed by each of these authorities. But truth is not relative. It's not "my version" of the truth or "your version" of the truth. Truth is conformance to reality. Truth is what is really real. And to find truth you have to look at the evidence for yourself.

There is no conflict between science and religion. Science is simply man's search for understanding the universe. Physics and faith go hand in hand. Biology and belief are compatible companions. Chemistry goes with Christ; geology reflects God. But there is war between the world views of naturalism and Christian theism.

There is no conflict between science and religion. But there is war between naturalism and Christian theism.

Naturalists believe that there is no God—that the entire universe and life is the result of chance. With naturalism the only authority is self. I decide what is important. I decide what is real. I decide what is moral. I decide what I want out of life—pleasure, power, prosperity. Because naturalists believe the entire universe is the result of chance they also believe that there is no purpose, no plan, no significance to the universe or life. Human life is animal life. At death you cease to exist. When you die you rot. And because the lifetime of individuals is short and the life-span of most cultures and nations is much longer, naturalists believe in-dividual humans are of little worth compared to nations and societies.

Christians believe the universe is created with plan and purpose, that human life is precious, that we are not our own—that our bodies belong to God because He created us. We believe what we do is significant, that humans have immortal spirits—we live forever. Christians believe we are more important than nations and societies. Nations will end. They will die out. We will not. Christians believe the moral law is built in to all people because we were made in the image of God. And not only do all people know the moral law, not only do they expect this behavior from others and claim it's "not fair" when someone violates this code, but all people break this moral law every day. We know what's right. We don't do it. And Christians have the only answer to this dilemma.

I grew up in a Christian family, believed in Jesus and was baptized in the Name of the Father, the Son, and the Holy Spirit at age 10. I had full faith in God, in the divinity of Jesus Christ, and in the truth of the Bible. But during high school, then at Stanford University and during medical school at UCSF my faith was severely tried. The prevailing cultural view promoted vigorously by my professors was that life started by chance billions of years ago (abiogenesis), that evolution supplied the explana-tion needed for the development of all forms of life including humans, and that it was unreasonable to believe in God.

As I trained to become a scientist with a degree in chemistry and then as a medical doctor it was hard to reconcile my training with my beliefs. My doubts became insistent. Was there any substantive evidence for God? Was there any good reason to believe that Jesus was God? And what about the Bible? Could it be trusted to be accurate and true?

As I dove deeply into the search for these answers I was amazed to find that the "evidence against God" was full of unproven assertions and prior beliefs. And I found that the evidence for God was overwhelming and convincing.

Christians need to understand the evidence that supports their beliefs. That is why this book was written. If you are not a Christian, or not a believer in God at all, I ask you to look at the evidence for yourself, rather than submitting to the opinion of popular authorities. As the evidence for God becomes persuasive you will have to consider this question: What should I do now? You could start by praying. The prayer you probably already know starts with "Our Father." It's a good way to start because He already knows you. You just have to get to know Him.

For every Christian, there is a prime directive—the guiding principle for all our thoughts and actions. Jesus was asked which was the most important commandment in the Law. It is recorded in Matthew 22:36-40. Jesus said, "'Love the Lord your God with all your heart and with all your soul and with all your mind.' This is the first and greatest commandment. And the second is like it: 'Love your neighbor as yourself.' All the Law and the Prophets hang on these two commandments."

As you consider the topics in this book don't forget the prime directive. In everything love God and love your neighbor as you love yourself. Who is your neighbor? Jesus answered this question with the parable of the Good Samaritan. Your neighbor is everyone you have an opportunity to help whether it is convenient for you or not. Jesus said "your neighbor" even includes "your enemy." This means your neighbor is everyone you have contact with and everyone your actions affect—at school, on the playing field, at work, at home, on the freeway, on social media. As you get to know God you will find many opportunities to love your neighbor.

CHAPTER 1

DOESN'T SCIENCE PROVE THERE IS NO GOD? WHY DO CHRISTIANS BELIEVE GOD CREATED THE UNIVERSE AND LIFE?

Many scientists claim that abiogenesis and evolution explain the origin of life. What scientific evidence indicates that life is designed—that God created life?

I am a scientist. I hold a degree in chemistry from Stanford University, a doctorate in medicine from University of California San Francisco, and advanced certification in surgery. Some have asked me, "How can you still believe in God? You are a scientist. Doesn't science prove there is no god?" As a scientist I can assure you that science certainly can not prove there is or is not a god. Anyone who claims that it can is in the realm of philosophy, not science. Origins are the realm of hypotheses and theories. These are guesses about how things started. No one can absolutely prove what started the universe and life. No one can prove god, and no one can disprove god. In other words, belief in god is not falsifiable by science. But there are lots of clues from scientific findings about the origin of the universe and life.

If you have already decided that the scientific evidence indicates there is no god, you are in for a surprise. When the science of forensics is applied to investigating the origin of life the conclusion is clear. Someone caused the origin of life—someone intelligent. It was not the result of chance. Forensics deals with probable cause, and forensics points directly to a creator.

Abiogenesis means life started from no life by chance, with no plan and no design. It is also called chemical evolution. This is the prevailing cultural belief in our time—the worldview of naturalism. This is the view supported by many high school textbooks and natural history museums. This theory states that the early earth had a reducing atmosphere that allowed formation of amino acids and other simple organic compounds that collected in the oceans as a "primordial soup" with the makings for life. From there the organic compounds concentrated at various locations (shorelines, oceanic vents etc.) and by multiple chance chemical interactions more complex organic polymers – and ultimately life – developed in the soup.

This theory is not supported by science. Even the simplest form of life—the first living cell—is not only staggeringly complex but contains a huge amount of information. Chance can not and does not produce information. The presence of information is one of the basic criteria used to detect intelligent origin. Cellular DNA and RNA are programs that produce living cells. They are code. Programs are information rich. Programs require a programmer. If there is no programmer the computer dictum applies—garbage in, garbage out. Chance is "garbage in."

DNA and RNA are programs—code. Programs require a programmer. If there is no programmer the computer dictum applies—garbage in, garbage out. Chance is garbage in.

There are two ways to look at the scientific evidence for the origin of life—"bottom up" and "top down." Bottom up means that you start by looking at the molecules that might have been available to start life and see if there is any evidence that they might have formed into the first living cell by natural undirected means—by chance—in the "primordial soup." This is the way most often used to describe how abiogenesis

18

might have happened. This is the way abiogenesis is described in Wikipedia. It assumes that amino acids in the prebiotic soup bonded together into proteins, that nucleic acids formed in the prebiotic soup and bonded together into RNA and DNA, and all this evolved into the first living cell.

Another way to look at the evidence is "top down." In my college math textbooks the questions were at the end of each chapter, and the answers were in an appendix at the end of the textbook. If you couldn't find a way to work the problem as it was given, you could start with the answer and work backwards, hoping to get a hint of how to solve the original problem. Starting with the problem is bottom up. Starting with the answer is top down. If you already know that the answer is life—the first living cell, you can start there and see if there is any reasonable way that this first form of life could happen by chance natural events. Top down means that you start by looking at the details of the first living cell —the simplest possible form of life—to see if there is any mechanism that could explain its occurrence by chance natural means.

The remainder of this chapter is a look at the scientific evidence regarding the origin of life. We will consider the evidence from both a bottom up and top down approach. If you don't want to work hard to understand the chemical and physical evidence, please skip to the chapter summary. But the crux of the argument for design is here in the biochemical evidence.

The biochemical evidence is key

To begin with you need to understand a few basics of the biochemistry of amino acids, proteins and nucleic acids. I will keep this as simple as possible. Amino acids are the building blocks of proteins. Proteins are made of chains of amino acids. Proteins are absolutely necessary for life. There is no life without proteins. They provide structure and function in every cell. So let's start with amino acids. An amino acid looks like this.

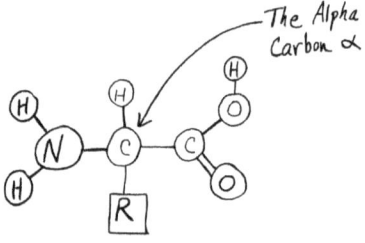

It can also be notated like this.

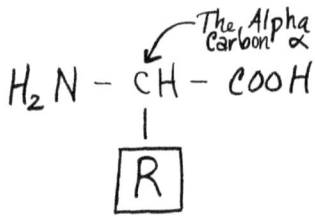

$$H_2N - \overset{}{C}H - COOH$$

The C's are carbon atoms; the H's hydrogen atoms, the N's nitrogen atoms; the O's are oxygen atoms. The line between them indicates a covalent chemical bond in which they are sharing two electrons. An amino acid consists of the amino end, the end with the nitrogen and two hydrogens notated as NH2, and a carboxyl end, the end with the carbon, two oxygens and a hydrogen indicated as COOH. The double lines between one of the oxygens and the carbon indicates a double bond in which the carbon and oxygen are sharing four electrons. Note that each carbon has four bonds, each oxygen has two bonds, each nitrogen has three bonds, and each hydrogen has a single bond. In the middle of each amino acid is the alpha carbon. It's called the alpha carbon because it is the first in the chain attached to the R group. The R group can consist of a single hydrogen atom (as in glycine) or many different atoms of carbon, oxygen, nitrogen, hydrogen, and sulfur.

Amino acids in solution (dissolved in water) are diagrammed slightly differently, because the two ends become electrically charged, negatively on the COOH end and positively on the NH2 end by attachment of the hydrogen proton (the hydrogen nucleus), minus its electron, to the amino end. This leaves the amino end positively charged and the carboxyl end negatively charged. Don't let this throw you. It's the same amino acid, but it looks slightly different in solution, like this. It is in solution that amino acids bond to one another.

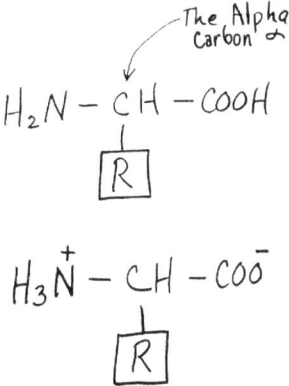

The Alpha Carbon α

$$H_2N - CH - COOH$$
$$| $$
$$\boxed{R}$$

$$H_3\overset{+}{N} - CH - CO\bar{O}$$
$$| $$
$$\boxed{R}$$

The top diagram shows the amino acid not in solution. The bottom diagram shows the same amino acid in solution.

There are 21 amino acids needed to support life. The difference between one amino acid and another is only the difference in the "R" group. This "R" group "side chain" determines the different chemical characteristics of each amino acid. The R group can be as simple as a single hydrogen attached to the alpha carbon (glycine) or very complicated. Each amino acid except glycine also has two racemic forms—mirror images—left and right handed. The left and right forms occur because the amino acids are three dimensional, like your hands. Here are diagrams of several of these amino acids.

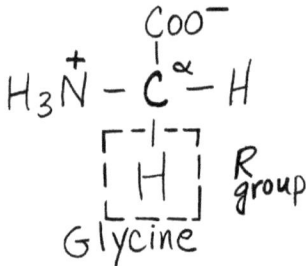

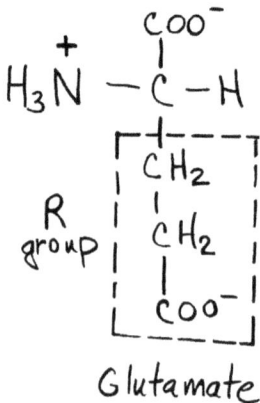

$$H_3\overset{+}{N} - \overset{\overset{\displaystyle COO^-}{|}}{\underset{\underset{\boxed{H}\ \ R\ group}{|}}{C^\alpha}} - H$$

Glycine

$$H_3\overset{+}{N} - \overset{\overset{\displaystyle COO^-}{|}}{\underset{\underset{\underset{\underset{COO^-}{|}}{CH_2}}{CH_2}}{C}} - H$$

R group

Glutamate

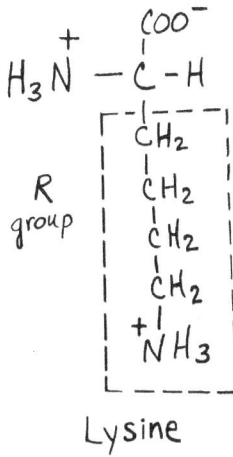

Lysine

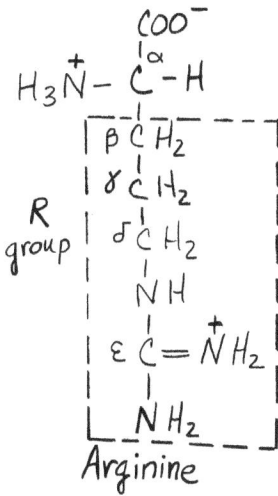

Arginine

$$\overset{\text{COO-}}{\underset{\displaystyle \beta \dot{C} H_2}{\overset{+}{H_3} \dot{N} - \overset{\alpha}{C} - H}}$$

R group

$\beta \dot{C} H_2$
$\gamma C H_2$
S
$\delta C H_2$

Methionine

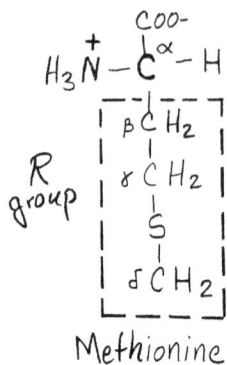

Note that the basic structure at the top is the same. The R groups make each amino acid different.

Proteins are built from amino acids bonded together at the alpha carbon by a peptide bond. A peptide bond looks like this.

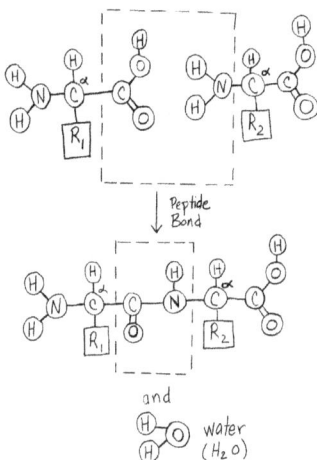

Peptide Bond

and

water (H_2O)

Note that the carboxyl group carbon bonds to the nitrogen of the adjacent amino group, and one molecule of water is produced. This still leaves one amino group at one end and a carboxyl group at the other end. You can see how additional peptide bonds at the alpha carbon at each end can produce long chains of amino acids. These are called polypeptides. Proteins are special polypeptides.

You can also see that there are additional amino groups and carboxyl groups in the R group side chains. If these form peptide bonds the function of the protein is destroyed. The way the protein folds into the specific shape needed for its function depends on the individual amino acids having peptide bonds only at the alpha carbon. If the polypeptides are in the right specific order, if they are all bonded only at the alpha carbon, and if all the amino acids are in their left handed racemic form, a protein results. Different proteins have different functions depending on the specific order of the amino acids.

Proteins are the building blocks of life. You can't have life without proteins. How are they produced so that they use only left handed amino acids, have only alpha bonds, and are put together in the right specific order? Ribosomes in the cell are protein factories following DNA and RNA instructions. Ribosomes are complex structures of multiple proteins and RNA strands. They "read" instructions from messenger RNA, then bond each amino acid in sequence at the alpha carbon, preventing any side chain peptide bonds.

DNA and RNA are chains of nucleic acids connected together with phosphodiester bonds. The structure is similar to proteins, but DNA and RNA use only right handed nucleic acid bases instead of left handed amino acids. Like amino acids in proteins, the specific order of the nucleic acids in DNA and RNA determines how they function. The DNA double helix is the spiral structure DNA forms with its long chains of nucleic acids. DNA is made up of multiple genes. Each gene codes for

construction of a specific protein, or a specific function in the cell. Therefore, DNA is the code for construction of the living cell.

Proteins are made up of amino acids
- alpha carbon peptide bonds
- left handed racemic form
- in specific order

DNA and RNA are made up of nucleic acids
- phosphodiester bonds
- right handed racemic form
- in specific order

Now with these basics let's consider the evidence. There are only three scientific experiments that support the possibility of "bottom up" abiogenesis—the Miller experiment, the Fox experiment, and the "RNA world" experiment. Remember, bottom up means trying to find some mechanism that could produce precursors to life like functional proteins, RNA, or DNA from the "primordial soup."

The first was the Stanley Miller experiment in 1952, in which he was able to produce the 21 amino acids needed for life from non-organic pre-cursors (simple chemicals like ammonia, water, methane, sulfur, and hydrogen). The second experiment that supported the possibility of abiogenesis was Sidney Fox's in 1958. He was able to show that purified amino acids would chemically bond (connect together) under certain high temperature conditions. Fox called these "proteinoids" (things like pro-teins, but not actual functional proteins). The proteinoids combined to form small globules that Fox called "microspheres." They formed clumps and chains reminiscent of bacteria.

In the Miller experiment, a closed system with boiling water produced steam. Ammonia, methane, sulfur, and hydrogen were added. To simu-late lightning strikes continuous electrical sparks were produced in a chamber containing these basic compounds. As a result, amino acids were produced by this "natural chance" means.

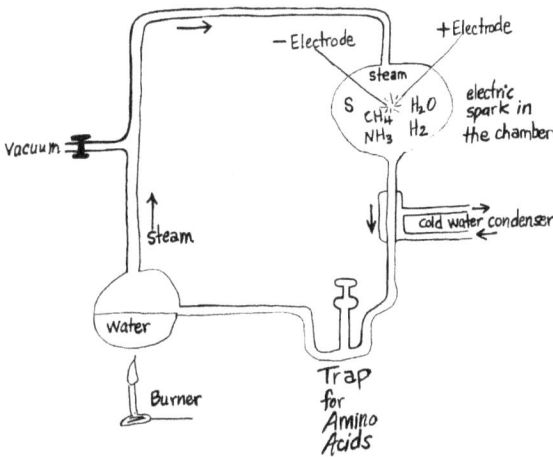

The Miller experiment produced these amino acids in mixed left and right handed *racemic* forms and randomly bonded together including bonds on the R group side chains. But living organisms use only left handed amino acids, and right handed nucleic acids (DNA and RNA). The amino acids produced in his experiment were not pure amino acids.

Many were chemically bonded on their side chains to other amino acids, making them unusable for production of proteins.

The initial reaction to these experiments was that if amino acids could be produced by chance natural means, and if they could chemically bond when heated then surely these amino acids could form the proteins needed for life by chance natural means. But this conclusion was short lived. Further experiments showed that to produce the kinds of proteins needed for life several additional conditions had to be met. First the proteins all had to be left handed racemic form. Could any natural conditions produce an abundance of only L-form amino acids? No. Second, for protein production the amino acids had to be bonded with the peptide bond only at the alpha carbon, with no peptide bonds on the side chains. Could any natural conditions produce only alpha peptide bonds between amino acids? No. Third, to produce proteins all the amino acids had to be bonded in a specific order. Since the natural tendency is for amino acids to bond randomly, could any natural conditions produce the specific order needed? No. Three strikes and you're out. No chance natural conditions could account for the production of proteins needed for life.

- Miller experiment 1952—it is possible for amino acids to form under chance natural conditions BUT all mixed up and stuck together
- Fox experiment 1958—purified amino acids under optimal conditions will bond BUT produced only random and useless polypeptides

"RNA World" is another "origin of life" hypothesis that tries to account for some way that life might have arisen by chance without any intelligent plan or design—without a program.   This is the third experiment said to support abiogenesis.   RNA world refers to production of self-replicating ribonucleic acid (RNA) molecules that might be precursors to a living cell. Observations about living cells point to the fact that proteins large enough to self-fold and have useful activities could only have come about after RNA was available to make the proteins in the first place, since ribosomes are critical to the manufacture of proteins.

Ribosomes are the protein factories that "read" the program for a specific protein from messenger RNA.   But the simplest ribosomes contain 42 separate proteins and 3 RNA chains. Messenger RNA (mRNA) is transcribed from a strand of DNA, giving this mRNA the specific instructions to build a protein.   Each individual amino acid needed for the protein is then picked up by a transfer RNA (tRNA) and brought to the ribosome, where the active site of the ribosome, reading the messenger RNA program, bonds each amino acid, in order, to the last amino acid in the growing protein chain with an alpha peptide bond, preventing any other chemical bonds between the adjacent R group amino acids.

Since ribosomes themselves are immensely complex containing multiple proteins, and since ribosomes are needed for protein synthesis, it is a circular argument—a fallacy—to   insist that ribosomes could be produced by chance chemical bonding.   It is   circular because ribosomes are known to be critical to protein manufacture. But since ribosomes are themselves composed of multiple proteins, where did the proteins for the ribosomes come from?   "RNA World" is an hypothesis unsupported by logic or experiments.

It is possible to create self-replicating RNA molecules in a lab.   It requires researchers to build an RNA chain about 142 nucleotides long from two smaller chains, then supply energy and substrates in a solution designed to prevent hydrolysis (breakdown of the RNA nucleotide bonds). This allows the RNA to replicate itself. The process itself is complex, requiring an intelligent researcher to build the RNA chain in the first place, then maintain the environment so replication can proceed. RNA precursor nucleotides never spontaneously form active RNA, because random bonding does not produce the specific type of bonds—phosphodiester bonds—or the specific order needed for active RNA.

RNA World assumes ribosomes came first, because you have to have ribosomes to make proteins. But ribosomes are made of proteins.

Objections to the RNA world hypothesis have arisen from many origin of life researchers. According to Wikipedia, "Molecular biologist's dream" is a phrase coined by researchers Gerald Joyce and Leslie Orgel to refer to the problem of emergence of self-replicating RNA molecules as a precursor to life on earth, as any movement towards an RNA world on a properly modeled prebiotic early earth would have been continuously suppressed by destructive reactions. They noted that many of the steps needed for the nucleotides formation do not proceed efficiently in prebiotic conditions. Joyce and Orgel specifically referred to the need for "a magic catalyst" that could "convert the activated nucleotides to a random ensemble of polynucleotide sequences, a subset of which had the ability to replicate".

Joyce and Orgel further argued that nucleotides cannot link unless there is some activation of the phosphate group, whereas the only effective activating groups for this are "totally implausible in any prebiotic scenario", particularly adenosine triphosphate. Joyce and Orgel reject "the myth of a self-replicating RNA molecule that arose de novo from a soup of random polynucleotides." In other words, RNA world is not a reasonable hypothesis for the origin of life by chance. It just doesn't work.

To summarize the 'bottom up" view of abiogenesis:

1. Proteins are needed for life. They are made of chains of amino acids. But amino acids never self-assemble into proteins. They must be put together by specific cellular machinery—ribosomes—that use DNA and RNA to provide the program. The ribosomes, themselves complex combinations of proteins and RNA, provide for all alpha bonding using only L-form amino acids and putting them in the specific order required.

2. DNA and RNA are needed for life. Nucleic acids do not self assemble into DNA or RNA. DNA and RNA are made of right handed nucleic acids arranged in a specific order. Like the order of amino acids in protein, this specific order is information, with genetic code similar to digital computer code. It is clear that the only source of information is intelligence. For production of useful code you need a programmer. Information, like computer code, does not occur by chance.

3. Finally, the RNA world hypothesis is not supported by scientific evidence. Only tightly regulated experiments with heavy manipulation from the researchers allow RNA replication in the lab outside a living cell. The evidence indicates that the RNA World hypothesis is unreasonable as pointed out by origin of life researchers Gerald Joyce and Leslie Orgel.

Bottom up evidence from the promordial soup cannot support abiogenesis.

Since bottom up evidence cannot support abiogenesis, how about top-down? Starting with the simplest living cell can we work backward to find a chance natural mechanism that would explain its origin?

We now have a good example of what the simplest living cell would look like. It is Syn3.0. This is the type of cell envisioned by those who believe in undirected evolution as the "first living cell," also called the "last universal ancestor." It would be the precursor of all living cells on earth. It would be the first cell at the base of the "tree of life" in evolution textbooks.

In May 2010, a team of scientists led by J. Craig Venter became the first to successfully create what was described as "synthetic life," which was based on the actual DNA of a simple bacterium. They started with a goat intestine bacterium called mycoplasma genitalium. Mycoplasma genitalium is a primitive bacterium that has one of the smallest genomes of any free-living organism in the world, with 525 genes. That's a fraction of the size of even another common bacterium like E. coli, which has 4,288 genes. M. genitalium's tiny genome made it the first target for Stanford and J. Craig Venter Institute researchers who wanted to copy the genome to make the first artificial living cell.

They synthesized its DNA molecule, containing its entire genome. Because m. genitalium reproduces very slowly, they introduced the synthesized DNA into two other similar bacterial cells that reproduced rapidly. The two bacterial cells that were suitable for this type of transplant of genetic material were mycoplasma mycoides and mycoplasma capricolum.   In each, they removed the native DNA, then transplanted the synthetic DNA copy of m. genitalium into these "DNA empty" cells. They left all of the cellular machinery in these recipient cells intact, including their ribosomes, their cell walls with ion pumps, structural and transport proteins, metabolic processes to use food to produce stored energy, and all the accompanying controls with feedback loops.

They named their new bacterium JCVI-syn1, Syn 1.0 for short. This single-celled organism contained four identifiers written into its DNA to label it as synthetic and to help trace its descendants. The identifiers included a code table for the entire English alphabet with punctuations, the names of 46 contributing scientists, three quotations, and the email address for the cell.

They let the new bacterium grow. Then, one by one, they looked at each and every gene and removed it. If the bacterium died, they put it back—it was necessary. If it wasn't, it stayed out. The first version of the synthetic genome, JCVI-syn1.0, had just over a million base pairs, and JCVI-syn2.0 had 576,000. On March 25, 2016 Venter reported the creation of JCVI-Syn 3.0, a synthetic genome having the fewest genes of any freely living organism (473 genes).

This cell is alive. It is the world's simplest living cell. It reproduces in its laboratory dish every three hours. "You cannot live without all but one or two of the genes in this genome," said Venter. A previous study published by the National Center for Biotechnology Information in 1995 suggested that a genome that coded the most basic life form would be at least 256 genes. Venter noted that "everybody was off—by a third." It took 473. The team said that many of the genes had unknown functions, but were nonetheless necessary for the organism to grow and replicate.

To summarize, the total number of genes required for the simplest living cell was 473. The DNA base pair count was 531,000, and all of the nucleic acid bases had to be in precisely the right order. During construction of Syn 3.0 one base pair error (one nucleic acid error in 531,000 nucleic acids in specific order) in one gene caused cell death. It took the research team three months to find and correct this error.

"Perhaps," Venter said, "life isn't built from independent parts, like something in a machine shop. Instead of focusing on genes, maybe we should consider the whole operating system—not the genes but the 'genome,' a functioning whole. Life may not be machine-like. We may be orchestral. Genes whisper. They amplify. They turn other genes on and off and make them louder, softer, even silent. They cause cascade effects, sending notes through the system that loop back, creating loops within loops. A living thing may be a constantly changing melody, orchestrated by its constantly changing parts."

After the original publication of this accomplishment, there were many comments posted by the scientific community. One of special interest was written by atheist Ron Braithwaite on 4/22/16. He noted, "The above findings have major significance for the origins of life. Given a 'perfect' biological soup—whatever that might be–the odds against an accidental biochemical combination to produce the very first gene must be long, indeed. The odds against the accidental production of dozens of quite different genes and then having them accidentally join together in the absolutely necessary arrangement, also must be mathematically enormous. Also, such an initial life form MUST have had mechanisms to protect its own nucleic acid with incredible perfection, otherwise its reproduction products [daughters] would not have had the minimal 'stuff' possessed by their own minimal, maternal 'cell'. These imperfect daughters would have been nonviable and immediately faded into the surrounding 'soup.' Now, given an early earth with an ideal biological soup containing all the building blocks necessary for that first life, there must have been an incredible number of failed 'life' experiments. Still, *as a materialist*, I must wonder about the probability of the whole thing. We are presented with the evidence that 'life' arose spontaneously at least once but *the numbers seem to be very much against it even given ideal circumstances*."

He was simply stating the obvious. Life cannot have started by chance in the primordial soup. Note the immense complexity. The simplest possible living cell has 473 genes. Each gene is 1000 or more DNA base pairs long. This cell has 15 genes for glucose transport, 14

34

for ribosome biogenesis, 10 for protein export, 9 for transcription, 7 for RNA metabolism, 5 for DNA topology (shape of DNA), 3 for chromosome segregation, 3 for DNA metabolism, 3 for protein folding, 89 for translation (DNA to mRNA), 35 for RNA (3 types, r, m, and t), 16 for DNA replication (copying DNA, making new), 21 for lipid salvage and biogenesis, 21 for cofactor transport and salvage, 12 for ribosomal RNA modification, 17 for transfer RNA modification, 7 for efflux (excretion of waste products), 19 for nucleotide salvage (reusing bases), 6 for DNA repair (fixing copying errors), 10 for metabolic processes (energy use), 31 for membrane transport (moving materials in & out of the cell), 4 for redox homeostasis (oxidation/ reduction processes), 10 for proteolysis (breaking down proteins), 9 for regulation, 1 for cell division, 15 for lipoproteins

The simplest living cell isn't simple
- It contains 473 genes
- It has 531,000 base pairs in its DNA
- All DNA bases must be right handed
- All DNA bases must be phosphodiester bonded
- All DNA base pairs have to be in the exact right order

(cell wall components), 2 for transport and catabolism of non glucose carbon sources, and 79 for unknown functions.

Note that every one of these 473 genes must have its DNA base pairs in exactly the right order. All these genes must be present for the simplest form of life. And even then these genes must be housed inside a functioning cell wall with ion transport capacity. Well, you might say, so what? Can't this all happen by chance given enough time? Why can't information occur by chance? Is there any scientific test you can apply that would let you determine whether something happened by chance, or whether it happened as the result of some action by an intelligent agent?

It turns out there is.

There is a scientific field, forensics, that focuses on causes—natural events versus those caused by some agent. It is important to distinguish the cause. With any event, object, or structure, we want to know: Did someone cause it to happen? Did it happen by accident? Did it have to happen?

Forensics is the science of determining causes
- Contingency
- Complexity
- Pattern

Entire industries are devoted to distinguishing between accident and design. These include intellectual property law, insurance claims investigation, forensic science for crime investigation, investigation of plagiarism and data falsification. Was this fire an accident or arson? Did this person die a natural death or was she murdered?

The science to determine what is accidental (natural) and what is designed (caused by an intelligent agent) was well defined by William A Dembski. He described this method in Science and Evidence for Design in the Universe. Three things are required to determine that something is produced by an intelligent agent: contingency, complexity, and specification (an independent, recognizable pattern.) These are the criteria for design.

Contingent means that what happens depends on something else, that it doesn't just happen on its own naturally. To show that an event is contingent one must first show that it is not the result of a natural law or algorithm. For example, a salt crystal is a complex structure that is the result of the laws of chemistry. When salt water evaporates salt crystals invariably form. By contrast, the order of scrabble pieces on a board is not the result of any natural laws. No laws dictate where each scrabble letter is placed. The order of scrabble pieces is thus contingent whereas the structure of the salt crystal is the result of chemical laws. The position silverware is set on a table is not the result of any natural laws. It is contingent. The configuration of ink written on a sheet of paper is not the result of the physics and chemistry of paper and ink. It is contingent. The specific sequencing of DNA bases is not the result of the bonding affinities between the bases. It is contingent. DNA and RNA bases can and will bond together in any order. The order has to be defined by something or someone.

So the position of scrabble pieces on a board, the way silverware is set, writing with ink on paper, and the order of DNA bases are all contingent. They all depend on something or someone putting them in order. Contingency is required for an event to be of intelligent origin.

Second, the event or item must be complex. Complexity is a form of probability. Complexity and probability vary inversely: the greater the complexity the smaller the probability. But complexity (improbability) is not enough to eliminate chance as the origin. If I shuffle a deck of cards and turn them over one by one and record the results, this is complex but meaningless.

The third thing you need to determine if something is of intelligent origin is a suitable pattern. Not just any pattern will do. If you shoot arrows at a large barn, then paint a target around each arrow so that the arrow sits in the bull's eye, this does make a recognizable pattern, but it is a pattern made up after the arrow was shot. It indicates that you hit the target every time, but it is a lie. Suppose instead that you nail a target to the barn wall, then shoot at it, and each arrow you shoot hits the target. Now we can tell how good you are at shooting arrows. The criterion of "suitable pattern" means an independently given pattern, a pattern you can recognize as a message or a picture or a structure.

To apply the algorithm to detect intelligent origin, you ask three questions. 1. Is it contingent (not the result of natural laws: salt crystal vs. scrabble pieces)? 2. Is it complex (improbable)? 3. Does it conform to an independently given pattern (not fabricated)?

You use these criteria every day without thinking. If you come upon a sand castle on the beach or a message inside a heart shape scratched into the sand "Kathy, I love you—Phil" you know instantly that this is not the result of chance. They are contingent—no law of wind and waves makes this happen. They are complex—grains of sand in special shapes. They fit a recognizable pattern. You know what they are—what they represent. And you know someone made them. You know they didn't just happen by chance.

A good example of how to use these criteria is in SETI, the government funded search for intelligent signals from outer space. SETI is the **S**earch for **E**xtra **T**errestrial **I**ntelligence. With the use of multiple large radio telescopes signals from space are analyzed for an intelligent source. This search is based on the hypothesis that if life arose naturally on earth from inorganic precursors by chance, then almost certainly there are other intelligent beings "out there" that might be communicating with electromagnetic signals similar to what we use. So far there have been no such signals received. However, there is a good hypothetical example of what such a signal might look like.

Carl Sagan was a popular scientist and a strong supporter of SETI. He used to begin his TV series "Cosmos" with the statement, "The universe is all there is, all there was, and all there ever will be." He was an atheist. Among his writings is a novel about how SETI gets an intelligent message from space that leads to "contact" with intelligent people from other worlds. In Carl Sagan's novel Contact, and in the movie made from that book, the fictional event that triggers SETI to determine that the signal is coming from an intelligent source is a set of prime numbers from 2 to 101 that continue repeating over and over. Prime numbers are numbers divisible only by 1 and themselves. Three things were satisfied to determine that the signal was coming from an intelligent source: contingency, complexity, and specification.

In the case of the prime number sequence in the novel Contact the pattern of zeros and ones forming the sequence of prime numbers was not the result of the laws of physics that govern electromagnetic waves transmitting these numbers. They could have been any numbers in any order. The sequence indicated contingency.

If the prime number sequence in Contact had been too short, say 2, 3, and 5 repeating (only three numbers) this could happen by chance. But the prime number sequence from 2 to 101 requires 1126 bits of information in specific order, a level of complexity that indicated intelligent origin.

Third, the prime number sequence was a recognizable independent pattern. Anyone who knew mathematics would recognize it instantly.

Thus, it is clear from the example of Carl Sagan's novel Contact that SETI uses these criteria, although not in a well-defined form, to determine intelligent origin.

Using these criteria for origin, natural chance versus intelligent design, applied to the DNA molecule of the simplest living cell, we note that

1. It is contingent. Its specific order is not the result of any physical or chemical laws. Experiments confirm that that there is no natural tendency for DNA bases to bond in any specific order. And nucleic acid bases bonded randomly have no function.
2. It is complex, with 531,000 base pairs in specific order
3. It conforms to an independently given pattern. It is the instructions for construction of a cell. It is the blueprints for a living cell.   It is a program. It is, in fact, not just "analogous to" a message. It IS a message, a set of plans and machinery for building a living cell.

Therefore, the DNA of the simplest living cell is not of natural origin. It is not the result of chance. It is the result of an intelligent designer, a programmer.

Forensics applied to DNA shows it is not the result of chance. It is designed.

To get some idea of just how much information is in the simplest living cell consider this—this book has far less information in it than the first living cell. If it is apparent to you that this book is the result of some intelligent agent writing it, and that it didn't happen by chance, it should be obvious that the DNA or RNA of a cell is the result of an intelligent designer.

A good example of how naturalists have tried to explain the origin of life is this. It is like the way I tried to solve one of the differential equations problems in my college math book. I started bottom up, with the problem, and worked on it as far as I could. It didn't get me to a solution. So I tried a different approach, top down, starting with the answer from the appendix at the end of the book, and working back to the original problem. I couldn't get that to work either. So I just put the work I had done in order on my paper, joining the two methods in the middle, and called it good, hoping my teaching assistant (TA) wouldn't notice that it didn't really work. When I got the paper back graded, there were check boxes by each step until the middle. At that point there was a lot of red ink, where my TA tried to figure out how I had gotten from where I was to the conclusion I was supposed to reach. There was a large red question mark in the margin at this step, and a comment in red. "Leap of faith?" She knew I was unable to solve the problem.

This is like the abiogenesis problem. The naturalist solution doesn't work. But the leap of faith is not a religious leap. It's a leap of faith to trust in naturalism—to trust that life could originate by chance when all the scientific evidence speaks otherwise. My math problem had a solution. But I wasn't heading the right direction. The origin of life has a solution too. But it isn't what the naturalists hoped it would be.

The conclusion is clear. Scientific evidence indicates that the first living cell was designed—created by an intelligence. Scientific evidence supports the conclusion that life did not start by chance abiogenesis, but by intelligent design.

There is an easy way to remember all this. Miller and Fox did the original experiments that made it appear that life could have started by chance. But real proteins require all **A**lpha bonds (A), all **L**eft handed amino acids (L), and all the amino acids bonded in **S**pecific **O**rder (SO). This makes the acronym **ALSO**. DNA is a **Spiral** structure. DNA is a set of instructions—a program—that is of intelligent origin (the product of a **mind**). How do we know that DNA is intelligent in origin, the product of a mind? We apply forensic scientific criteria used by **SETI** to the DNA molecule. The result is that the DNA spiral is intelligent in origin. How to remember **SETI**? Look at the middle two letters. ET. That's the name of the Extra Terrestrial in the movie "ET"!

This little ditty makes it easy to remember:
Abiogenesis rocks
With Miller and Fox
Amino acids and peptides
Creation mocks

But add in what's needed
For life, to find
ALSO and Spiral
The blueprints from mind

How do we know this?
From science it's clear
SETI to Spiral
Says, "Intelligence here!"

EVOLUTION OF SPECIES

In addition to the scientific evidence that leads to the conclusion that the origin of life was planned and executed by an intelligent agent, the scientific evidence confirms that undirected evolution cannot account for the development of higher life forms from lower life forms.

Evidence is strong that all living things have much in common. It is easy and convincing to demonstrate that humans are closely related to chimpanzees. But the fact that humans share 99% of the DNA of chimpanzees does not mean that we descended from them or from a common ancestor by random mutation as evolution requires. If an inventor is creating new items she often uses previous models because they work, and improves them by adding new information. This is not "directed evolution." The changes needed for new species must be input all at once, because random changes one base pair at a time cannot account for the huge number of changes needed for the origin of new species.

Species do adapt to various environmental conditions. The Galapagos finch beaks demonstrate this, becoming thicker in drought to crack tougher seeds, thinner in wet years. Mutations in malaria confer drug resistance to anti-malarial agents, and then revert back to normal when the drugs are no longer used.

But the "theory of evolution" with no causative agent, that is, evolution by chance, requires that small incremental changes in DNA—random changes from mutation—*must* account for the development of new species—new animal and plant forms. There are several problems with this. First, the vast majority of mutations are harmful to the organism. Second, it is now clear that new animal and plant species require hundreds of changes in the DNA all at once to allow new proteins, new protein folding patterns, new binding sites on these proteins, and changes in the operating systems of the cell (the developmental gene regulatory networks—dGRN's—that work like complex electrical control circuits). These are not incremental random changes. These are changes on the scale of building a new three story office building, fully functional, and having it happen by chance. Third, new species occur suddenly in the fossil record with no intermediate forms. There is no naturalistic explanation for this sudden onset of new species. There is a name for it—punctuated equilibrium—but no explanation of how this could happen by chance random mutation.

Incremental random mutation of DNA cannot explain new species

It is clear that chance cannot account for the origin of life, and chance random incremental mutations cannot account for the development of higher life forms (animals and plants) from lower life forms. Abiogenesis and unplanned evolution are simply not supported by the scientific evidence. This conclusion is based on molecular biology, mutation rates and mechanisms (in animals, malaria, bacteria, and HIV), longitudinal genetics studies, paleontology (the fossil record), study of dGRN's, ENCODE Project findings, thermodynamics (the energy required for production and coding of living organisms), and forensics (study of the clues— how did this happen, what is the most likely explanation?). For more detail please consult Michael J. Behe's book, The Edge of Evolution: The Search for the Limits of Darwinism.

Two additional books that detail these scientific findings are Darwin's Doubt, by Stephen C. Meyer; and Evolution: Still a Theory in Crisis, by Michael Denton.

The resistance to accepting the scientific evidence for intelligence as the cause of the origin of life and the development of species is not based on science but on preconceived belief. This belief is best illustrated by the statement that "despite the scientific evidence that indicates plan and design, there 'must be' some way that natural processes alone can produce life, and higher species from lower species."

This belief is naturalism, a religious belief that there is no god and everything that has happened "must have" happened by accidental natural causes. The prophets of this religion are Richard Dawkins (The God Delusion), Daniel Dennett (Breaking the Spell, Religion as a Natural

Phenomenon), Michael Martin (<u>Atheism: A Philosophical Justification</u>), and Carl Sagan (<u>Cosmos</u> "the Universe is all there is, all there was, and all there ever will be.") Their holy book is Charles Darwin's <u>On The Origin of Species</u>.

Chapter Summary

Whether you look at the possibility of abiogenesis from the bottom up with life arising from a "primordial soup" or from the top down looking at the simplest possible living cell the conclusion from the scientific evidence is compelling. The origin of life is clearly the result of intelligent plan and design, not chance. The application of forensic science to the DNA molecule makes this plain.

To detect intelligent design using forensic science, three criteria are used.

1. Contingency. Is it something that depends on something else (contingent)? Or is it something that always happens naturally? Is it like salt crystals forming from evaporating salt water (not contingent), or is it more like scrabble tiles (contingent), that have to be placed in a specific order and orientation to make words?

2. Complexity. Is it something that is likely to happen naturally? Or is it something with hundreds (or thousands) of bytes of data (information)?

3. Independent pattern. Can you tell what it is? Is it a message? A picture? A set of plans? Or is it a random pattern, with no discernible meaning?

Using these criteria for origin, natural versus intelligent, applied to the DNA molecule of the simplest living cell we note that—

1. It is contingent. Its specific order is not the result of any physical or chemical laws. Experiments confirm that that there is no natural tendency for DNA bases to bond in any specific order. If nucleic acid bases are allowed to bond randomly, they have no function.
2. It is complex with 531,000 base pairs in specific order
3. It conforms to an independently given pattern. It is the instructions for construction of a cell. It is the blueprints for a living cell. It is, in fact, not just "analogous to" a message. It IS a message, a set of plans and machinery for building a living cell. It is a program.

Therefore it is not of natural origin. It is not the result of chance. It is the result of an intelligent source.

Science as taught in public schools proposes a body of dogmatic truth. It is the belief system of naturalism—there is no god and everything in the universe came about by chance. Abiogenesis—the belief that life originated by chance in the primordial soup—is taught as fact. Evolution—the belief that all species developed by chance by random incremental mutation from the first living cell acted upon by natural selection—it taught as fact.

But scientific evidence does not support either of these assertions. The only evidence for chance origin of life is that amino acids can be produced by natural methods from inorganic precursors in a reducing atmosphere. This was established by Stanley Miller in 1952. But all subsequent experiments aimed at producing life by chance have failed. Why? Because the simplest possible form of life—the first living cell—is not simple. This was established in 2016 by J. Craig Venter and his team at Stanford. The simplest possible living cell has 473 genes and 531,000 base pairs of DNA all in precise sequential order. Nothing simpler than this is life. Nothing simpler can survive under natural conditions. And this level of complexity cannot be the result of chance.

How do we know this? We apply the scientific criteria of forensics. The conclusion is clear. Life is designed. Similarly, when the hundreds of changes in DNA base pairs needed for a new species is noted, there is no way this can happen by chance random mutation of DNA. Adaptation of species does occur. This is clear from studies of finch beaks in the Galapagos. But new species requiring hundreds of DNA changes are not possible by chance mutation.

Science is clear. Life and species did not develop by chance. But scientific evidence is ignored in public schools because the belief system of naturalism prevails.

So don't be persuaded by the constant assertion by naturalists that abiogenesis and evolution can explain everything. The weight of Western culture is overwhelmingly in favor of abiogenesis and evolution, because naturalism is the dominant religion in most of our institutions of higher learning. But the weight of scientific evidence, evidence that life is designed—not accidental—will prevail, because it is true.

CHAPTER 2

IS THERE ANY ACTUAL EVIDENCE FOR THE EXISTENCE OF GOD?

There are four kinds of evidence that lead to the conclusion that God is real. What are they?

The first is physical evidence from science. The second is study of the human mind and behavior. Third is the written record in the Bible, the history of how God has interacted with humans over many centuries. Finally, there is the evidence of experience, how God has changed lives and answered prayers.

First, the physical evidence

The physical evidence from science comes in five categories—the "**5 CLUES**."

The first letter in CLUES is the C. It stands for the "**C**osmic Welcome Mat" or **C**onstants, the precise constants of physics that determine the way in which matter and energy behave. For the universe to exist and for life to exist these 15 constants must have precise values. These constants include the speed of light c, the gravitational constant G, the strong and weak nuclear forces, the elementary charge e, the fine structure constant that characterizes the strength of the electromagnetic interaction (alpha), the rest mass of quarks, the cosmological constant lambda, and several others. If any of these constants deviated even slightly from the values they now hold, the universe and life would not be possible. The probability of this happening by chance is so small as to make it essentially impossible. The universe, therefore, appears to be "fine tuned" for intelligent life. This is called the Anthropic Principle.

In addition, the cosmic welcome mat includes all the special features of earth that allow life to exist. In the 1960's and 70's it was believed that there were likely to be many worlds in the universe that could support life. With billions of stars and trillions of planets there just "had to be" some others that could support life. But as the actual conditions required to support life were investigated over the ensuing years earth was seen to be a very special planet. It is not just that the sun is the right kind of star and that the earth is the right distance from the sun to be in the "goldilocks zone" for the right temperature. Another requirement for life is the precise mass of our planet. If it were slightly larger methane and ammonia gas would remain close to the surface of earth. These are toxic. And if the mass were slightly less water vapor would dissipate into space leaving us waterless. If earth rotated slightly slower the temperature differences between night and day would be deadly. If it rotated faster it would produce extremely high winds. Other features include the presence of a large planet in our solar system, Jupiter, that protects the earth from most comet strikes, and a moon precisely the right size to allow tides and stabilize earth's rotational axis (allowing seasonal changes). For a more detailed explanation of the many special features of earth, things that make it likely that it is unique in the universe, see Eric Metaxas' book Miracles.

One explanation for the "cosmic welcome mat" is that a divine creator caused the fine tuning in an intentional way. Many scientists have observed that the appearance of design in the universe is overwhelming. Dr. John Lennox, professor of mathematics at Oxford University, observed, "The more we get to know about our universe, the more the hypothesis that there is a creator God, who destined the universe for a purpose, gains in credibility as the best explanation for why we are here."

But the alternate explanation from naturalism is that it actually did happen by chance and the appearance of design is only "appearance" and the universe is not really designed. Faced with the massive scientific evidence that the "cosmic welcome mat" is not a chance event, those who insist that there is no god, that chance "must be" the explanation for everything, propose an alternate theory. This is the multiverse theory—that ours is only one universe of many in trillions of alternate universes existing side by side, but ours is the only one we experience. And if there are an infinite number of alternate universes with an infinite number of chances for getting all the criteria just right then it is certain that chance would produce this universe. This is called the "Many Worlds in One" interpretation of quantum mechanics and string theory. There is no scientific evidence for the MWO theory. It cannot be proved or disproved (falsified). MWO is a belief not testable by science. Sir John Polking-horne, an eminent quantum physicist, put it this way, "Let us recognize these speculations for what they are. They are not physics, but in the strictest sense, metaphysics. There is no purely scientific reason to believe in an ensemble of universes." Richard Swinburne, Emeritus Professor of Philosophy at Oxford University, put it bluntly. "To postulate a trillion-trillion other universes, rather than one God, in order to explain the orderliness of our universe, seems the height of irrationality." Dr. Edward Robert Harrison, an astrophysicist and cosmologist, said, "The fine tuning of the universe provides prima facie evidence of deistic design. Take

"To postulate a trillion-trillion other universes, rather than one God, in order to explain the orderliness of our universe, seems the height of irrationality."

Richard Swineburne, Oxford University

your choice: blind chance that requires multitudes of universes or design that requires only one.   Many scientists, when they admit their views, incline toward the design argument."

Those who continue to insist that "the universe is all there is" and that "god is a delusion" are our culture's scientific version of Pharisees. Think about it. The Pharisees were opposed to Jesus because His ideas and teachings about what was true and who God was upset their preconceived notions.   They refused to believe the evidence that was right in front of their eyes.   In the same way, the atheist scientists and philosophers of our generation are Pharisees, refusing to look at the evidence, instead clinging furiously to their belief that there is no god. But pursuing the truth is not a new idea.   Shakespeare nailed it in The Merchant of Venice. "The truth will out."

The second letter is L, and it stands for "Life." The question is "How did life start?"   The scientific evidence is clear.   Life did not start by chance or by purely naturalistic causes. Life originated from intelligence. Life originated with information. As detailed in Chapter 1 whether abiogenesis is considered from the "bottom up" from amino acids and nucleic acids or from the "top down" by looking at the simplest living cell, the scientific evidence is strongly in favor of design. There is no source for the information in DNA that is basic to life except an intelligent agent. Information in a program requires a programmer.

The third letter is U for Universe.   There is an interesting YouTube video by Gerald Schroeder, PhD, Physics, that explains what the Big Bang is all about.   You can find it at https://youtu.be/eQVm8RokoBA. What is the origin of the universe? It's not clear even yet that the Big Bang is the final answer although there is lots of evidence that points to it, like the expansion of the universe and the background radiation of the universe. Before the Big Bang theory was accepted the generally accepted scientific view was that the universe was eternal—it had always existed. And this scientific view of the universe conflicted with the Bible because the Bible said, "In the beginning God created the heavens and the earth." With the arrival of evidence for the Big Bang including the expanding universe and the residual background radiation of the universe it confirmed that the Bible was right after all. The universe had a beginning. It was *not* always there.

The universe was not always there. The Big Bang theory supports the Bible. "In the beginning, God created the heavens and the earth."

Then the big question became, "How did the universe start?" As Dr. Schroeder explains, what is clear from physics and mathematics, especially particle physics and string theory, is that something *can* be created from nothing as long as natural laws are in effect like quantum mechanics and general relativity. Considering particle physics it is possible to separate antimatter from matter to create something from nothing. It is possible with string theory that the universe was created from nothing. If the natural laws that allowed the formation of the universe from nothing were in effect before the Big Bang, before the universe existed, then these natural laws existed before the universe—they predated the universe. They were not part of space-time. They were not "physical." They had creative power—power to make the universe out of nothing. Dr. Schroeder finds this very interesting. Something that is not physical, acts on the physical, can create the physical, predates the universe, and is outside of time and space—is the cause of creation—the cause of the Big Bang. He points out that there is a person who fits this definition. "In the beginning, God created the heavens and the earth." This fits the evidence and nothing else does.

The fourth letter is E for Entropy. Entropy is one of the "laws" of physical chemistry. Entropy means that "everything runs down." It's called "time's arrow." It is not reversible. Entropy means that in any closed system (like the universe itself), things slowly deteriorate, become more random, more diffuse, less organized. Low entropy means more useable energy and more organization. High entropy means less useable energy, less organization. Low entropy is good. One of the ways to describe increasing entropy is this—you can't unscramble an egg. Anything that becomes more organized like a growing living thing does it by breaking down something else and using that energy to grow more complicated, so that overall there is greater entropy (more disorganization, less useful energy in the whole system) at the end. If you walk this back to the beginning you would say, "In the beginning was something immensely energetic and organized." And you would be right.

The odds of our universe's low entropy condition (high useful energy, high degree of organization) obtaining by chance alone are on the order of 1 in 10 in 10 to the 123rd power, an inconceivably low number.

Roger Penrose, Oxford University

Our universe has amazingly low entropy. Roger Penrose of Oxford University calculated that the odds of our universe's low entropy condition obtaining by chance alone are on the order of 1 in 10 in 10 to the 123rd power, an inconceivably low number. Understand this. This is one in ten to a one with 123 zeros. 10 to the 123 power is a huge number, more than the total number of atoms in the entire universe. But 10 to the 10 to the 123—an impossibly large number. So the inverse of this, one chance out of this number in the denominator is simply impossible by chance. Roger Penrose is not a believer in God. He interprets his calculations by saying that this low entropy condition could not happen by chance, but there must be some non-god cause. His theory is that the universe reinvents itself by Conformal Cyclic Cosmology (CCC). He believes that as all matter and energy is finally sucked into black holes that entropy will actually be reversed, that "time's arrow" will run backwards, because the degrees of freedom in a black hole are less than outside the black hole. He postulates that as the final massive black holes eventually evaporate by Hawking radiation that a new big bang may occur, with echoes of the old universe still present in the new universe.

The explanation that fits the scientific facts best is this: the low entropy (highly organized with lots of useful energy) condition of our universe is intentional. Living organisms have very low entropy. They are highly complex, highly organized, and information rich. They have both low thermal entropy and low organizational entropy (ability to generate work from metabolism, high information content). This low entropy condition with extreme specified complexity and information content can only occur with an intelligent origin.

The fifth letter is S, for Strings. A popular mathematically based theory for the makeup of matter and energy is string theory. String theory predicts that all matter and energy is created by vibrating strings. The length and vibration of each one dimensional string (no mass) produces specific sub atomic particles. This theory requires at least 10 dimensions and in some versions requires 13 dimensions. One way to visualize this is like seeing a hose stretched out over the Grand Canyon. From a distance it looks one dimensional like a string, but on close examination it has at least two more dimensions, width and height of the outside and inside of the hose. Then if you looked at each part of the hose you could see even more divisions each with its own dimensions. So with string theory one dimensional strings in loops vibrate creating the elementary particles of matter (quarks, leptons, antiquarks, antileptons) and forces (gauge bosons and the Higgs boson). One dimensional means they have no "thickness" only length. This is why it's called "string theory." There is little physical evidence to support this theory. The observed universe has only four dimensions including length, width, height, and time. There are competing theories, less popular, like twistor theory. But if string theory proves to be correct what makes up the matter and forces in the universe is strings in motion. This is like music, the tones representing each type of particle and force. You could reasonably conclude from string theory that the Big Bang was a loud song by an amazingly powerful Singer.

Second, the Human Evidence

All this so far is physical evidence, evidence from science. But we have another way to examine whether God exists and that is us—humans—what we are like. This is the line of evidence C.S. Lewis followed in his book, Mere Christianity. There are multiple aspects of our psychology and behavior that point to God's existence. One way to remember them is this. **M**ale **A**nd **F**emale **H**e **C**reated them—**MAFHC**.

The first letter stands for **M**orals. Humans are moral creatures in an otherwise amoral universe. The rest of the universe lives by the law of survival of the fittest with no "right or wrong" entering into the picture. Humans have an inbuilt belief in a universal moral law—a moral obligation—the Tau, and there is no satisfactory purely natural explanation for it. Why anything should be "right" or "wrong" is dependent on this inbuilt law.

Humans are moral creatures in an amoral universe. Morality protects and nourishes the weak and strong alike. But natural selection is indifferent—it weeds out the weak and unfit.

Christians expect this moral behavior in humans because they believe that humans were "created in the image of God." But if there really is no god, if we are just the product of an uncaring unplanned natural process there is absolutely no reason to expect that humans would have a sense of moral obligation. It's clear that most atheists have a fine-tuned sense of morality. Christians expect this since we believe all humans are created in the image of God with inbuilt moral laws. What atheists don't have is any good explanation for why they have these moral beliefs.

Friedrich Nietzsche tried to explain the origin of morals in his famous book, On the Genealogy of Morality (1887). He claimed (as do all naturalists, who refuse to believe in God) that morals are due to the evolution of the human brain and that morals are the result of evolution acting on our larger sentient brains. He believed that evolution produced morality and ethics as an advantage to survival. He proposed that morals are relative to your culture and personal situation and that the strong (or the majority of the strong) decide what is moral. But when you examine Nietzsche it is clear that his conclusions are nothing more than assumptions and circular reasoning. He called the Biblical law "master-slave morality" that came from Israel in bondage in Egypt. There is absolutely no scientific or historical evidence that would support Nietzsche's claims.

Christians believe morals come from God and morals are absolute. They believe morality does the opposite of natural selection (survival of the fittest.) Morality is non-selective—it protects and nourishes the weak and the strong alike. The opposite is the struggle of life with survival of the fittest. Natural selection is indifferent—it weeds out the weak and unfit. Since survival of the fittest is the dominant feature of evolution the emergence of morality (protecting the weak and unfit) would be a destructive mutation that would soon become extinct. If humans believe morals are important, if morals are "hardwired" into the human psyche this is strong evidence for God

The second letter **A** stands for **A**rt. This category fits the acronym **BALM**, the fact that humans appreciate **B**eauty, **A**rt, **L**ove and **M**usic. If naturalism is true then beauty, art, love and music are nothing but biochemical responses to help us survive. In the experience of beauty, art, love and music almost all humans feel there is real meaning in life, joy in music, and that love means everything. Humans believe that beauty, art, love and music make life significant. This is powerful evidence for God.

The third letter **F** is for **F**ree Will. Most people believe they have free will. Our society certainly thinks so. Our laws are designed to protect us and keep order. A good example is traffic lights. We all know red lights mean stop and green lights mean go. And the reason we have our police give tickets to those who run red lights is because we know that each driver has the free will to stop in accordance with the law or go through in violation of the law. The reason we feel justified in calling red light running "unlawful" is because we are sure that each driver has the choice to stop or not and if they decide to do wrong they have to pay. On the other hand, if we are truly the product of an uncaring universe we are creatures of chance and our lives have no significance. If it is true that our behavior is determined only by our response to our environment there is no such thing as free will. Thinking atheists agree with this conclusion. In a closed system (the universe) if there is no god everything is determined and free will is an illusion. Even those philosophers most in favor of the naturalist position agree with this assessment. But the rest of us do believe in free will. And it is powerful evidence for God.

The fourth letter **H** stands for "**H**ardwired belief in God." Human behaviorists have found conclusively that belief in God is "hardwired" in the human brain. There is an inbuilt tendency for humans to believe in God. The Naturalist argument against this is called the "clue killer" argument. The "clue killer" argument goes like this—if naturalism is true then everything about humans is only the result of natural selection. These inbuilt beliefs about morals, free will, love, and God are not actually true—they just help us survive. But if the "clue killer" argument were true, if these human characteristics were not real facts about ourselves but were only delusions that conferred survival advantage then *all* human beliefs are subject to this same delusion including the atheist's belief that there is no god and that everything happened by chance.

This leads us to the fifth letter **C**. It stands for human **C**ognition—reason. If you can't trust your reason to tell you the truth about morals and free will you can't trust your reason for anything. But we do trust our reason. We all trust logic. So it is reasonable to trust that our innate belief in God is rational and true. We know that morals are important. We believe in free will. We love music and art. We believe that what we do is significant. We know there is a God. And all these human characteristics are powerful evidence for God.

Third, the Bible Evidence

The third line of evidence for God is the Bible. Christians believe the Bible is accurate and true. It tells us who God is, what He expects of us and how we can come into right relationship to Him. Christians believe the Bible is historically true and that the miracles and events described actually happened. Some parts of the Bible are clearly poetry such as "The Song of Solomon" and meant to be understood as poetry. But the life of Jesus as described by his contemporaries writing in the Bible New Testament is the actual historical record recorded by eyewitnesses. The claim by non-believers—that the gospels (Matthew, Mark, Luke, and John) were written hundreds of years later, that they were recollections retold and embellished with stories of miracles and resurrection from the dead—these claims have no historical basis. When critics try to refute the claim that the Bible is true they start by eliminating all the parts of the Bible where miracles are described because to them "miracles are impossible." But when the actual historical evidence is examined it is clear that these gospel accounts were recorded by eyewitnesses, within the lifetime of those telling the story. The most reliable historical records are eyewitness accounts, and the fact that the Bible New Testament is eyewitness accounts is supported by strong historical evidence.

Fourth, the Evidence from Experience

Christians believe in God because of their experience. Not only does the scientific evidence support intelligent design, not only do human characteristics indicate we are "made in the image of God," not only does the Bible ring true as historically correct, but we also have the experience of God. Those who have made the decision to "follow Jesus," to believe in Him, have experienced the relief that follows the confession and forgiveness of sin, the peace that comes from knowing the love of God, the sense of belonging as a child of God, the knowledge and direction that comes from the indwelling Holy Spirit—the part of God that lives in us and assures us of His reality and presence. It all comes together in belief. Father loves us. Jesus shows us who the Father is. Jesus said, "If you have seen Me, you have seen the Father." We know the character of Father. He is Love and we love Him back.

Chapter Summary

There are four types of evidence for God.
1. Physical Scientific Evidence: The 5 CLUES—the **C**osmic welcome mat, the origin of **L**ife, the origin of the **U**niverse, **E**ntropy, and **S**tring theory—all support the existence of God.
2. Human Evidence: The MAFHC group, **M**orals, **A**rt (Beauty, Art, Love, Music), **F**ree will, **H**ardwired belief in God, and **C**ognition are evidence that we were created by God.
3. The Bible New Testament is eyewitness accounts, and the Bible can be trusted.
4. Christians have experienced the Living God, and can testify to His reality and presence.

There is abundant evidence to warrant belief in God. But knowing the evidence is not enough. You have to make the decision for yourself—to surrender your will to Jesus and follow Him as your Lord. When you are sorry for your sins, when you ask Jesus to take over your life the Holy Spirit will find His home in you. As you put your whole trust in Jesus you will stop worrying because God has everything under control. Nothing can happen that He doesn't already know about. He will give you a way out of temptations. He will give you peace in trials. He promises abundant life. And He delivers on His promises.

CHAPTER 3

HOW CAN GOD ALLOW EVIL AND SUFFERING?

How can a good God, who is all powerful and all knowing, allow evil and suffering? What are 5 reasons God can be Good, Omniscient, and Omnipotent, and still allow suffering and evil? Good = always interested in doing things that benefit us. Omniscient (all knowing) = knows everything and every time. Omnipotent (all powerful) = has complete power and control over everything.

One of the primary arguments used against the existence of God is the argument from suffering and evil. Many people refuse to believe in a God who allows suffering and evil to exist. When natural disasters like tsunamis, earthquakes, tornadoes, and floods cause widespread loss of life and suffering the news stories are full of opinions about how if there was a God He would certainly not allow this. In the face of famine and war many people turn away from God in disbelief. They just can't imagine that God would allow such suffering and evil.

We need to face this question. If God is good, if God is all powerful, if God knows everything, why would He allow such suffering and evil? There are at least five good answers to this question.

1. Free Will entails suffering and evil. "Entails" means that suffering and evil are "inevitable consequences" of free will. Most people believe we humans actually have free will, the ability to make choices about what we decide to do. Those who take Naturalism seriously, the belief that there is no god and that the big bang and evolution explain everything, have to admit that if all the world and all its creatures are the product of chance with no purpose or plan then every-

thing is already determined by our environment and free will is just an illusion.

But people do not believe this! In all human societies there are rules to follow and punishments meted out for failure to follow these rules. We know that red traffic lights mean "stop." There is a decision we must make each time we come to a red light. We can choose to stop or choose to go right on through (free will.) We authorize our police officers to enforce the red light rule. When they see someone go through a red light they don't say, "Well, they did that because they were trained by their environment to break the rule so it's ok." No, what they do is pursue that person, pull them over, write a ticket and make them pay a fine. And if that person repeats the offense too often they will spend time in jail to remind them that their society expects them to follow this rule, to exercise their free will in such a way that they do not injure others.

If we have free will—the freedom to think and choose—some will choose to do evil. Thus free will entails evil. "Entails" means that it is an unavoidable consequence of the condition. It is clear that God felt that giving us free will was important enough to allow us to suffer its consequences including pain and evil. As said by CS Lewis in Mere Christianity, "Why, then, did God give them free will? Because free will, though it makes evil possible, is also the only thing that makes possible any love or goodness or joy worth having."

Think about it this way. At a certain age parents allow their children the freedom to make choices. The choice the child makes may be against the parents' wishes and against their beliefs. The choice may cause suffering for the child or others. The parent loves the child and hopes for the child's success. But giving them choice is part of growing up and the pain of learning the consequences of wrong decisions is a necessary part of letting them mature to responsible adults.

2. There is often a reason for suffering. A good example is the story of Joseph. It is told in Genesis, Chapters 37 to 50. Joseph was sold into slavery by jealous brothers. He was taken to Egypt where he served faithfully but was falsely accused of trying to rape his master's wife after he refused her sexual advances. He was thrown into prison for this. Later he was called to Pharaoh to interpret Pharaoh's dream. He correctly predicted from that dream 7 years of plenty followed by 7 years of famine. Pharaoh made him second in command in all of Egypt to prepare for the famine. Joseph became powerful and famous. When his brothers who had sold him into slavery came to Egypt to escape famine Joseph provided food and a new home for his family. He saved his family and all Egypt from starvation.

    Everyone suffers. It's part of life. Sometimes it looks like the suffering is pointless and destructive. But most of the time looking back on the experience we can see the benefit in our lives.

3. Evil is an argument FOR God, not against God. If there is no God the natural law that applies to all living beings is survival of the fittest. Evolution entails natural selection. In the absence of God Hitler's behavior was not evil but normal. He was powerful and well fit for his position in society. He convinced the majority of his countrymen that what he was doing was right. Only if you believe in some absolute moral standard can you call anything "evil." We don't call animals who kill other animals, even their own species, "evil." But we do call child molesters "evil." Humans are moral animals in an amoral world. Why? Because we were created in the image of God so we know when something violates God's moral order. The fact that we recognize "evil" is a powerful argument for God.

4. Suffering is something God did too—Jesus on the cross. Jesus claimed to be God and proved it with miracles and rising from the dead. His life, death and resurrection changed the world forever. And eyewitnesses to his life documented that Jesus suffered. He was in agony in the garden of Gethsemane on the night he was betrayed. He suffered as He was beaten and had a crown of thorns pressed into His scalp. He suffered as His hands and feet were nailed to the cross. He suffered as he struggled to breathe hanging on the cross. If God suffered too we know that suffering is not something that happens because God doesn't care. He cares. He is good. He suffered and died for you.

5. Heaven is a place where all suffering is reversed and all evil accounted for. Jesus promised that God would make a home for us where we would be together with Him. He promised that all wrongs would be avenged by God at the Last Judgment. "Vengeance is Mine, says the Lord." Deuteronomy 32:35. It's also in Romans 12:19. This is what lets us "turn the other cheek" and bear the evil meted out to us by others, repaying evil with kindness. We don't have to avenge the evil done to us and our loved ones. God tells us in unequivocal terms not to do this. See Matthew 5: 38-45. God will be sure all evil is accounted for and we will have our suffering turned into joy as we arrive in heaven.

Here is a memory aid indicating the reasons God can be good, omnipotent, and omniscient, and still allow suffering and evil.

People try to say that God cannot be Good, Omnipotent, and Omniscient because
- If God is Good and Omnipotent and allows evil, then He must not know about it.

- If God is Omnipotent and Omniscient and allows evil, then He must not be Good.
- If God is Good and Omniscient and allows evil, then He must not be Omnipotent.

  Therefore, no such God.

They are trying to turn GOD into GOO (Good, Omnipotent, Omniscient, can't be all three and still allow evil)

The answer: a **FRESH** bottle of **GOO Gone**:

F R E S H

- **F** = **F**ree will entails suffering and evil
- **R** = **R**eason for suffering. Bible example: Joseph and his coat of many colors
- **E** = **E**vil is an argument for God, not against God
- **S** = God **S**uffered too, Jesus on the cross, so it can't be that God doesn't care
- **H** = **H**eaven is a place where all suffering will be reversed

Chapter Summary:

A good, powerful, and all knowing God can allow suffering and evil for at least 5 reasons.

1. Free will entails suffering and evil. God gave us freedom to choose and some choose evil.
2. There is often a reason for suffering. The classical Bible story that illustrates this is the story of Joseph sold into slavery in Egypt.
3. Evil is an argument FOR God, not against God. If there is no God there is no reason to call anything evil.
4. God suffered too. Jesus suffered in the garden and died on the cross. If God allowed this it can't be that He doesn't care about our suffering.

5. Heaven, our reward, is a place where all suffering and evil will be reversed.

The next time you encounter a news story that claims that God can't exist because of some natural disaster they are reporting, or the next time you hear someone say that they just can't believe in God because of all the suffering and evil in the world, remember a **FRESH** bottle of Goo Gone. Remind yourself and that person that free will is important and that if God were to eliminate all suffering and evil, He would have to eliminate our freedom as well. Explain free will using the example of the way parents give their children the freedom to make decisions on their own, knowing they will certainly make some poor decisions that harm themselves and others.

Learn to explain and defend the statement that "Evil is an argument FOR God, not AGAINST God."

Finally, understand that because of heaven, because God will correct all wrong and avenge all evil we do not have to respond to evil by trying to take revenge into our own hands. We can "turn the other cheek" knowing that it will come out right in the end.

CHAPTER 4

CAN YOU TRUST THE BIBLE?

What are five reasons Christians believe the Bible is true and can be trusted?

The Bible is under attack. A movement called "Historical Bible Criticism" labeled much of the Bible as unhistorical and untrustworthy. This was based on these critics' belief that the New Testament was written hundreds of years after the events, that miracles were impossible, and that the Bible was copied over and over from one language into another over centuries so that the copies must be riddled with errors, omissions, and mistranslations. The "Jesus Seminar" held in Napa, California 1985-1999 concluded that no more than 20% of Jesus' sayings and actions in the Bible were true. All His miracles were dismissed as "scientifically impossible." All of the prophesies in the Old Testament that described in detail the events of the life and death of Jesus were rejected as truth by the "Jesus Seminar" on the basis that these "prophesies" must have been written much later, after the life of Christ, and inserted into the older accounts since it was impossible that anyone could predict the future with such accuracy.

Historical Bible Criticism and The Jesus Seminar are not the result of scholars who believed in God. They were atheists even though they wore the collars of priests, rabbis, and ministers.

Understand clearly that "Historical Bible Criticism" and "The Jesus Seminar" were not the result of scholars who believed in God. They were Naturalists. They did not believe in God, did not believe Jesus could be God, did not believe miracles were possible. They were atheists even though they wore the collars of priests, rabbis, and ministers claiming to be experts in religion. They were the Sadducees of our time —"religious" but not believers in God.

Some claimed that alternate accounts of the life of Jesus, such as the Gospel of Thomas or the Gospel of Judas were more accurate than the Gospels in the Bible—Matthew, Mark, Luke, and John—but were suppressed by the Church because they did not confirm Jesus' claims to be God. Author Dan Brown claimed in his "novel" The Da Vinci Code that Jesus was a great teacher, but completely human, who hundreds of years after His death was made into a resurrected God by church leaders who wanted to gain status in the Roman Empire. Even though his book was a novel (fiction) the author made the claim that these things were historically true.

What is historically verifiable is that the entire Bible New Testament was written in the lifetimes of the witnesses recording the events, and was based on eyewitness accounts of the life of Jesus and the growth of the early Christian Church. Finding the Dead Sea Scrolls confirmed that there were no significant copying and translation errors, and that current copies of the Bible were unchanged from texts written two thousand years ago. Alternate "gospels" such as the Gospel of Thomas and the Gospel of Judas were first written more than 100 years after the four Gospels of the Bible New Testament were in widespread use. These alternate gospels—Judas and Thomas—reflect "gnostic" ideas such as

The Bible New Testament is eyewitness accounts, written within the lifetime of the witnesses to the actual events.

the belief that the body is evil, matter is evil, and only the soul and light are good. These things the Bible flatly contradicts.

As far as the claim that Jesus was made into a resurrected god hundreds of years later to gain status, Paul's letter to the Philippians written no later than 20 years after the crucifixion of Christ (AD 50) makes it clear that Christians were already worshiping Jesus Christ as Lord (see Philippians chapter 2.) All four New Testament gospels—Matthew, Mark, Luke, and John—were written within 40-60 years after Jesus' resurrection.

Some critics claim that the Bible supports slavery and subjugation of women. This has no support in the Bible. Slavery of New Testament times was not chattel slavery where the person was kidnapped and sold into slavery as the American, English, and Middle Eastern slave trade was conducted. With the type of slavery described in the New Testament only the person's labor was owned, not the person himself, and a person could earn or buy freedom from this obligation. The Bible condemns kidnapping and trafficking in slaves. (See Deuteronomy 24:7 and I Timothy 1:9-11.) It was Christians who led the fight to end slavery in England and in America. With regard to women, the Bible makes it clear that women and men have different roles, but not different worth.

There are good reasons why Christians trust the Bible to be true, accurate, historical and verifiable.

1. The Bible New Testament is eyewitness history. It describes Jesus' life and the start of the Christian Church. It was all written within the lifetime of these eyewitnesses.

A. The Gospel of Matthew was written by Matthew the Apostle, the former tax collector. He was an eyewitness of the actual events.
B. The Gospel of Mark was written by John Mark, disciple of the Apostle Simon Peter. The Gospel of Mark is Peter's eyewitness history.
C. The Gospel of Luke was written by Luke the Physician, close friend of Mary the mother of Jesus. The Gospel of Luke is the description of Mary's eyewitness account of the events.
D. The Gospel of John was written by the Apostle John, John the Elder, "the disciple whom Jesus loved." It is his eyewitness account of the life of Jesus.
E. Acts was written by Luke the Physician, based on his own personal observations as he traveled with the disciples of Jesus as the "good news" spread.
F. Romans and many of the other New Testament letters were written by Paul, a first person witness to Jesus. Paul was struck blind on the road to Damascus on his way to arrest Jesus' disciples. He heard Jesus speak to him and was converted to the faith.
G. James was written by the half-brother of Jesus, James, head of the Jerusalem church.
H. First and Second Peter were written by Simon Peter, Jesus' apostle, based on his personal experiences with the events of Jesus' life.
I. Revelation was written by the Apostle John when he was in exile from the Roman territories on the Island of Patmos. It is his account of what God showed to him.

2. The Bible is accurate. Modern Bibles compared to Dead Sea Scrolls (2000 years old) show absolute accuracy. Critics of the Bible claim that the Bible can't be accurate since it had to be copied by hand by scribes every hundred years or so as the parchment scrolls aged. With that many times copied there had to be errors so that current Bibles "must be" riddled with copying errors. But when the 941 different texts of the Qumran Cave Scrolls (Dead Sea Scrolls) were discovered in 1946-56 and were examined and compared to current Bibles written in Hebrew, Aramaic, and Greek, there were no substantial errors found. There are nearly complete copies of the book of Isaiah in the Dead Sea Scrolls that agree word for word with no substantial errors compared to modern copies.

3. The vast majority of the Bible was written in only two languages, Hebrew and Greek, and modern Bibles are translated directly from these original languages. In addition there are a few parts of the Bible written in Aramaic, an ancient language with similar roots as Hebrew. Bible critics claim that as the Bible was translated from the original Hebrew Old Testament and the Greek New Testament, and from there to the ancient English of King James' time, and from there to many different versions of the Bible that the original meaning was distorted beyond recognition and because of this the Bible can't be trusted. However, most of the Bible was written in only two languages, Hebrew and Greek, and modern translations are done directly from the original Hebrew, Aramaic and Greek. Hebrew and Greek are two "live" languages still spoken and used daily. So there are no distortions in the translations beyond that of trying to make meanings clear as we do when we speak any language foreign to us.

4. Fulfilled prophesy. There are 48 major predictions about the coming Messiah in the Bible Old Testament, many unlikely, all of which Jesus fulfilled. There is no historical evidence that any of this was faked, entered into the text later after the events of Jesus' life were known. The odds that anyone could fulfill all 48 major predictions by chance are so high as to be impossible. Yet Jesus, and only Jesus, fulfilled them all.

5. Jesus claimed to be divine and what happened after his death and resurrection confirmed these claims. If he had been only human we would expect the same result that we see in other people who claim to be God. We either lock them up in insane asylums or let them wander in their delusions as their followers drift away. But none of these had the same outcome that Jesus did. He rose from the dead and this fact was witnessed by more than 500 observers, most of whom were still alive at the time the New Testament was written in 50-100 AD. His Church grew from a handful of disciples to millions of believers who testified to changes in their lives that came from trusting in Him. It is clear from the eyewitness accounts of his life, death, and resurrection that Jesus was not crazy even though He claimed to be God.

Here is a way to remember these 5 reasons Christians believe the Bible is true and can be trusted.

The BIBLE song
(to the tune of "The B-I-B-L-E, Yes that's the Book for me, I stand alone on the Word of God, the B-I-B-L-E")

1. The B I B L E
Eyewitness history
They walked with Jesus, heard Him teach

The B I B L E

2. The B I B L E
With proven prophesy
Jesus' birth, His life and death
The B I B L E

3. The B I B L E
It's accurate, you see
The Dead Sea Scrolls read the same today
The B I B L E

4. The B I B L E
Hebrew and Greek the key
Just these two to modern speech
The B I B L E

5. The B I B L E
Jesus divine is He
He claimed He's God, He's not crazy
The B I B L E

5b. The B I B L E
Apostles there were three
Who saw Him shining as the sun
The B I B L E

5c. The B I B L E
Christ Jesus died for me
Then came alive, 500 saw
The B I B L E

Well-researched books about the Bible include:

Jesus and the Eyewitnesses, Richard Bauckham, 2006.

Jesus and the Victory of God, N.T. Wright, 1998

The Resurrection of the Son of God, N.T. Wright, 2003

The Historical Reliability of the Gospels, C. Blomberg, 1987

The Historical Reliability of John's Gospel, C. Blomberg, 2002

The New Testament Documents: Are They Reliable?   F.F. Bruce, 2003

The Historical Christ and the Jesus of Faith, C. Stephen Evans, 1996

Warranted Christian Belief, "Two (or More) Kinds of Scripture Scholarship", Alvin Plantinga, 2002

Is The New Testament History?, Paul Barnett, 1986

- All say the Gospel of Mark was written in the 70's AD, Matthew and Luke in the 80's, and John in the 90's, when many of the apostles and other eyewitnesses were still alive. See Luke 1:1-4. This is 40 to 60 years after Jesus' crucifixion. It was at a time when many of his opponents, officials, and bystanders were still alive.
- Mark names some of the eyewitnesses in Mark 15:21.
- Paul's letters were written 15 to 25 years after Jesus' crucifixion. He refers to eyewitnesses, five hundred at once who saw the risen Christ in I Corinthians 15:1-6.
- Paul commented to King Agrippa in Acts 26:26 that the crucifixion of Jesus was public knowledge.   He could not have made these statements about the death and resurrection of Jesus if they were not supported by hundreds of eyewitnesses in Jerusalem.   Christianity would never have gotten off the ground if these things were not well known.
- An analysis of the names of the people in the Gospel stories reflects the types of names Jews had before the destruction of Jerusalem in

70AD, and not the sharply different kinds of names the Jews had after 70AD when the Jews were scattered.

- The literary form of the Bible Gospels is too detailed to be anything but eyewitness accounts. CS Lewis said, "I have been reading poems, romances, vision literature, legends, and myths all my life. I know what they are like. I know none of them are like this (the Bible Gospels). Of this text there are only two possible views. Either this is reportage (eyewitness accounts)..or else some unknown ancient writer...without known predecessors or successors, suddenly anticipated the whole technique of modern novelistic, realistic narrative." This is from C.S. Lewis, Christian Reflections, 1967, p. 155. Modern fiction is realistic. It contains details and dialogue and reads like an eyewitness account. This genre of fiction, however, only developed in the last 300 years. In ancient fiction personal details are spare, and added only to promote character development or drive the plot. But in the Bible we are told about Jesus being asleep in a boat in a storm in Mark 4, Peter seeing Jesus on the beach when the boat was still 100 yards away, Peter catching 153 fish then jumping out of the boat to swim to shore in John 21, Jesus doodling with his finger in the sand when he was hearing the story of the woman caught in adultery in John 7. None of these things has anything to do with character development or plot. But they are typical of the things an eyewitness would record as a reporter.

- Many people refuse to accept the Bible because they believe it is "culturally outmoded and regressive," because it seems to support slavery and the subjugation of women. This is not true. Slavery of Bible times was very different than the chattel slavery of the American South. In the time of the New Testament, the slaves' labor was owned, but not the person. They were usually not poor. They made the same wage as free laborers. They could buy their freedom. African slavery was race-based, for life, based on kidnapped persons from Africa. The Bible condemns kidnapping and trafficking in slaves in Deuteronomy 24:7 and 1 Timothy 1:9-11. Christians were the first to declare that this kind of slavery was wrong and abolished slavery in England and the

United States. With regard to women it is clear from the Bible that in the Lord, there is no slave or free, no male or female. See Galatians 3:28. For more on this see Rodney Stark, For the Glory of God, 2004.

Although the Bible can be trusted for its historical accuracy, you must be careful to find out what it actually says, rather than relying on cultural interpretations. There are two good examples of this—the Christmas story and the six days of creation.

At Christmas we see displays and hear carols about Mary and Joseph traveling to Bethlehem, being refused a room at the inn, being referred to the barn or cave in back with the animals, baby Jesus lying in a manger, surrounded by barn animals, the wise men (Magi) opening their gifts to baby Jesus, shepherds just outside this circle adoring the new infant. But this is not the historical record in the Bible. Mary and Joseph did travel to Bethlehem to be registered for taxation. But the "inn" is not supported by the text. The actual translation is "spare room." Family houses in that time usually had two rooms. The big room was where the family and the animals lived. The animal section opened to the door and was about two or three feet lower in height than the family section. Cut into the upper level were feeding troughs for the animals—mangers. The family lived in the higher section of the main room. Many homes also had a "spare room" with its own door, for extended family members or visitors. In this case, Bethlehem was already crowded with visitors there for registration, so the "spare room" was already full. In this case, the visitors would have to stay with the family in the main room, without the privacy of their own room and door. This is the most likely scenario for the birth of Jesus. Mary and Joseph were staying with relatives in one of these houses and had to stay in the main room. It was natural for Jesus to be bedded in one of the mangers in this main room. This is how the shepherds found them.

It was about a year or two later when the Magi arrived. There is nothing in the Bible about camels. And when they arrived Jesus and his family were living in a house. Don't forget that the Magi brought valuable gifts. Mary and Joseph were not poor after this.

As for the six days of creation, a careful reading of the Hebrew makes it clear that six 24 hour days is only one of the possible interpretations. Another reading, completely compatible with a literal historical interpretation of the text, is that creation took place over millions of years. This is explained in Chapter 12.

More about why the Bible can be trusted can be found in Lee Strobel's book The Case For Christ, "Examining the Record."

Chapter Summary

1. The Bible is eyewitness testimony. This is the best historical evidence there is. The New Testament was written by eyewitnesses during their lifetimes.
2. The Bible Old Testament has 48 major prophecies about Jesus Christ in the Old Testament, hundreds of years before His birth. All were fulfilled by Jesus. Claims that these prophecies were written after the birth of Jesus have no evidence to back them.
3. The Bible is accurate despite multiple copies made over the last 2000 years. This is proven by examination of the Qumran Cave Scrolls (Dead Sea Scrolls) and comparing them to modern Bible texts.
4. The Bible was written in only two languages—Hebrew and Greek—both spoken today. Modern versions of the Bible are translated directly from the Hebrew and Greek.

5. Eyewitness accounts confirm that Jesus claimed to be God. This leaves only two possibilities. He was insane or He really was God. His life and the amazing results of His life—the Church—demonstrate that he was not crazy.

This memory aid may help you remember these 5 reasons you can trust the Bible.

"We see that Hebrew and Greek in the Dead Sea Scrolls prophesies Jesus"

- "We see" = eyewitness accounts
- "Hebrew and Greek" = only two live languages, translated directly into modern Bibles
- "Dead Sea Scrolls" = accurate copies, compared to texts 2000 years old
- "Prophesies" = over 48 major predictions about Jesus in the Old Testament fulfilled by His life, death, and resurrection
- "Jesus" = claimed to be God, not crazy. Events after His resurrection bear out His claims.

What does all this mean? It means you can trust the Bible to tell you real history. God created you for His pleasure. He loves you and has a plan for your life. He wants you to believe in Him, to love and trust Him.

The Bible is the "Owners Manual" for humans, letting us know how to live in such a way that we are in harmony with God and our neighbors. It makes clear that we are responsible for each other's welfare and the protection of our environment. It tells us what to do to make our lives peaceful and happy and what we should avoid.

Because the Bible is more than an ancient book of myths, because it really is "The Word of God" given by inspiration of God's Holy Spirit, it means you have a decision to make. If God is real and has made a way for you to know Him, it means there is no hiding from Him or pretending He is not there. You know the moral law. You know you break this law. Your life is open to Him. He knows you. Will you choose to know Him, and surrender your will to His?

CHAPTER 5

WHO IS JESUS?

What did Jesus say and do that leads Christians to believe He claimed to be God? Give at least five examples.

1. John 14:6. "I am the Way, the Truth, and the Life. No one comes to the Father except through Me."

2. John 14:9. "He who has seen Me has seen the Father"

- Scripture reference for these first two examples is John 14:5-11: Thomas said to Him, "Lord, we do not know where You are going, and how can we know the way?" Jesus said to him, *"I am the way, the truth, and the life. No one comes to the Father except through Me. If you had known Me, you would have known My Father also; and from now on you know Him and have seen Him."* Philip said to Him, "Lord, show us the Father, and it is sufficient for us." Jesus said to him, "Have I been with you so long, and yet you have not known Me, Philip? *He who has seen Me has seen the Father;* so how can you say, 'Show us the Father?' Do you not believe that I am in the Father, and the Father in Me? The words that I speak to you I do not speak on My own authority; but the Father who dwells in Me does the works. Believe Me that I am in the Father and the Father in Me, or else believe Me for the sake of the works themselves."

3. John 10:30. "I and My Father are one."

- John 10:27-30. "My sheep hear My voice, and I know them, and they follow Me. And I give them eternal life, and they shall never perish; neither shall anyone snatch them out of My hand. My Father, who has given them to Me, is greater than all; and no one is able to snatch them out of my Father's hand. *I and My Father are one.*"

4. Jesus forgave sin

- Luke 7:36-50 is the story of the sinner woman at the Pharisee's dinner who washed Jesus' feet with her tears, wiped them with her hair, kissed His feet and anointed them with fragrant oil. Jesus said, "Therefore I say to you, her sins, which are many, are forgiven, for she loved much. But to whom little is forgiven, the same loves little." Then *He said to her, "Your sins are forgiven."* And those who sat at the table with Him began to say to themselves, "Who is this who even forgives sins?" Then He said to the woman, "Your faith has saved you. Go in peace."

- Mark 2:4-12. And when they could not come near Him because of the crowd, they uncovered the roof where He was. So when they had broken through, they let down the bed on which the paralytic was lying. When Jesus saw their faith, *He said to the paralytic, "Son, your sins are forgiven you."* And some of the scribes were sitting there and reasoning in their hearts, "Why does this Man speak blasphemies like this? Who can forgive sins but God alone?" But immediately, when Jesus perceived in His spirit that they reasoned thus within themselves, He said to them, "Why do you reason about these things in your hearts? Which is easier to say to the paralytic, 'Your sins are forgiven you,' or to say 'Arise, take up your bed and walk?' But that you may know that the *Son of Man has power on earth to forgive sins*"—He said to the paralytic, "I say to you, arise, take up your bed, and go to your house." Immediately he arose, took up the bed, and went out in the presence of them all, so that all were amazed and glorified God, saying, "We never saw anything like this!" (this same story is told in Luke 5:18-26, and Matthew 9:2-8)

5. Jesus claimed to be sinless.

- John 8:46. Speaking to Jews who did not believe Him, he said, "Which of you convicts Me of sin? And if I tell the truth, why do you not believe me?"
- John 8:29. "The Father has not left Me alone, for I always do those things that please Him."

6. Jesus affirmed that He was the Messiah. "Tell us if you are the Christ!" "I am."

- John 4:3-26 is the story of the Samaritan woman Jesus met at the well. John 4:25-26. The woman said to Him, "I know that Messiah is coming" (who is called Christ), "When He comes, He will tell us all things." Jesus said to her, "*I who speak to you am He.*"

- At Jesus' trial before the Sanhedrin, he admitted that He was the Son of God. Luke 22:70-71. "Then they all said, "Are You then the Son of God?" So He said to them, *"You rightly say that I am."* And they said, "What further testimony do we need?   For we have heard it ourselves from His own mouth." Matthew recorded the same event in Matthew 26:63-65.   But Jesus kept silent.   And the high priest answered and said to Him, "I put You under oath by the living God: Tell us if You are the Christ, the Son of God." Jesus said to him, "*It is as you said.* Nevertheless, I say to you, hereafter you will see the Son of Man sitting at the right hand of the Power, and coming on the clouds of heaven." Then the high priest tore his clothes, saying, "He has spoken blasphemy! What further need do we have of witnesses?" The same event is described in Mark 14: 61-63. Again the high priest asked Him, saying to Him, "*Are You the Christ, the Son of the Blessed?*" *Jesus said, "I am.* And you will see the Son of Man sitting at the right hand of the Power, and coming with the clouds of heaven." Then the high priest tore his clothes and said, "What further need do we have of witnesses? You have heard the blasphemy."

- Jesus was talking with the blind man he had healed in John 9:35-38. Jesus heard that they had cast him out; and when He had found him, He said to him, "Do you believe in the Son of God?" He answered and said, "Who is He, Lord, that I may believe in Him?" And Jesus said to him, "You have both seen Him and it is He who is talking with you." Then he said, "Lord, I believe!" And he worshiped Him.

7. "Before Abraham was, I AM." Jesus claimed to be older than Abraham, and claimed the name of God for Himself, the name "I AM".

- John 8:56-59. "Your father Abraham rejoiced to see My day, and he saw it and was glad." Then the Jews said to Him, "You are not yet fifty years old, and have You seen Abraham?" Jesus said to them, "Most assuredly, I say to you, before Abraham was, I AM." Then they took up stones to throw at Him; but Jesus hid Himself and went out of the temple, going through the midst of them, and so passed by."

8. Jesus was seen in a glorified body (transfigured) conversing with Moses and Elijah, by 3 witnesses, James, John, and Peter.

- Matthew 17:1-9. Jesus took Peter, James, and John his brother, led them up on a high mountain by themselves; and He was transfigured before them. His face shone like the sun, and His clothes became as white as the light. And behold, Moses and Elijah appeared to them, talking with Him. Then Peter answered and said to Jesus, "Lord, it is good for us to be here; if You wish, let us make here three tabernacles; one for You, one for Moses, and one for Elijah." While he was still speaking, behold, a bright cloud overshadowed them; and suddenly a voice came out of the cloud, saying, "This is My beloved Son, in whom I am well pleased. Hear Him!" And when the disciples heard it, they fell on their faces and were greatly afraid. But Jesus came and touched them and said, "Arise, and do not be afraid." When they had lifted up their eyes, they saw no one but Jesus only. Now as they came down from the mountain, Jesus commanded them, saying, "Tell the vision to no one until the Son of Man is risen from the dead."

- This same event is recorded in Mark 9:1-10, Luke 9:28-36, and Second Peter 1:16-18 where the Apostle Simon Peter notes specifically that he was an eyewitness to this event.

9. Jesus rose from the dead with a living body, able to eat and be touched. After Jesus' death, burial, and resurrection he was seen alive by more than 500 people on at least 12 occasions. Most of these witnesses saw Him in the first 40 days after His resurrection before He ascended into heaven.

- Matthew 28: 1-10. Appeared to Mary Magdalene and the other Mary. Also recorded in Mark 16:9-11, and John 20:1-18.

- Luke 24:34. Appeared to Simon Peter. Also recorded in I Corinthians 15:5.

- Luke 24:13-35. Appeared to Cleopas and another disciple. This is also recorded in Mark 16:12-13.

- John 20-19-23. Appeared to 10 of the Apostles (Thomas was not there).

- John 20-24-28. Appeared to the remaining 11 Apostles including Thomas (Judas had betrayed Jesus, and was not there). "Then He said to Thomas, "Reach your finger here, and look at My hands; and reach your hand here, and put it into My side. Do not be unbelieving, but believing."

- The same episode is described in Luke 24:36-43. In addition, He told them He was not just a spirit, that He had flesh and bones and could eat food. "Now as they said these things, Jesus Himself stood in the midst of them, and said to them, "Peace to you." But they were terrified and frightened, and supposed they had seen a spirit. And He said to them, "Why are you troubled? And why do doubts arise in your hearts? Behold My hands and My feet, that it is I Myself. Handle Me and see, for a spirit does not have flesh and bones as you see I have." When He had said this, He showed them His hands and His feet. But while they still did not believe for joy, and marveled, He said to them, "Have you any food here?" So they gave Him a piece of a broiled fish and some honeycomb. And He took it and ate in their presence. Mark 16:14 also describes this event.

- John 21:1-24. Jesus appeared at the Sea of Galilee to Simon Peter, Thomas the twin, Nathanael of Cana, the sons of Zebedee (James and John) and two other disciples. He told them to cast their nets on the other side of the boat, and they caught a "multitude" of fish (153 fish). Jesus was cooking fish and bread over a fire of coals as they came to land.

- Matthew 28:16-20. Jesus appeared to the eleven disciples on a mountain in Galilee that He had indicated to them. Mark 16:14-18 is another report of this meeting.

- I Corinthians 15:7. Jesus was seen alive by James

- Luke 24:50. Jesus was seen by His disciples before His ascension into heaven. Acts 1:1-11 and Mark 16:19 record this as well.

- I Corinthians 15:1-8 is one of the earliest writings of the Bible New Testament, written 15-20 years after Jesus' crucifixion. It is Paul's summary of the events following Jesus' death, burial, and resurrection, and confirms that Jesus' followers were clear about their belief that Jesus was God. "For I delivered to you first of all that which I also received: that Christ died for our sins according to the Scriptures, and that He was buried, and that He rose again the third day according to the Scriptures, and that He was seen by Cephas (Peter) then by the twelve. After that He was seen by over five hundred brethren at once, of whom the greater part remain to the present, but some have fallen asleep. After that He was seen by James, then by all the apostles. Then last of all He was seen by me also, as by one born out of due time. For I am the least of the apostles, who am not worthy to be called an apostle, because I persecuted the church of God."

- Acts 9:1-19. Saul (later his name was changed to Paul) was on the road to Damascus to persecute the Church when Jesus appeared to him. Additional references to this are Acts 22: 1-16 and Acts 26:12-18.

There are multiple other eyewitness references to Jesus' claims to be God. These include the following:
- He said He was the only begotten Son of God. John 3:16
- He said he was authorized to give life to whoever He wanted. John 5:21
- He said that everyone should honor Him "just as they honor the Father. He who does not honor the Son does not honor the father who sent him." John 5:23
- He said that God gave Him authority to judge. John 5:26
- He said he was the "bread of life." John 6:35
- He said He came down from heaven. John 6:36
- He said that everyone who saw Him and believed in Him would have everlasting life; and He would raise him up on the last day. John 6:40 John 6:47
- He said that no one had seen the Father, except for Him; He had seen the Father. John 6:46

- He said, praying to His Father, "And this is eternal life, that they may know You, the only true God, and Jesus Christ whom you have sent." John 17:3
- He said, "And now, O Father, glorify Me together with Yourself, with the glory which I had with You before the world was." John 17:5
- He said to His Father in prayer, "You loved Me before the foundation of the world." John 17:24
- He said to Pilate, "You say rightly that I am a king. For this cause I was born, and for this cause I have come into the world, that I should bear witness to the truth." John 18:37
- He said he was the living bread which came down from heaven, and that "if anyone eats of this bread, he will live forever; and the bread that I shall give is My flesh, which I shall give for the life of the world." John 6:51
- He said, "I told you that you would die in your sins; if you do not believe that I AM, you will indeed die in your sins." John 8:24
- He said he was Lord even of the Sabbath. Matthew 12:8, Mark 2:28, Luke 6:5
- His followers confirmed with Him that He was the Messiah, God's Christ. Matthew 16:16, Mark 8:27, Luke 9:20
- He predicted His death, and that He would rise from the dead. Matthew 17:9, 17:23, 20:19, Matthew 26:32, Mark 9:31, Mark 10:34, Luke 9:22, Luke 18:33
- He said he would sit on the throne of His glory, with His Father, God. Matthew 19:28 Mark 13:26
- He affirmed that the crowd was right, crying out to Him as Messiah, "Hosanna to the Son of David." Matthew 21:16, Luke 19:38
- He said that He was the one who sent the prophets of the Old Testament. Matthew 23:34
- He said, "All things that the Father has are Mine." John 16:15
- He said, "I came forth from the Father and have come into the world. Again, I leave the world and go to the Father." John 16:28

It is perfectly clear from the Bible eyewitness accounts that Jesus claimed to be God. It is also plain that He acted rationally and spoke coherently with no signs of insanity.   The worldwide influence of His followers after His death and resurrection confirm these claims.

Here is an easy way to remember five examples of what Jesus said and did that leads Christians to believe that He claimed to be God.

The "Jesus Loves Me" Song

(to the tune of "Jesus Loves Me this I know, for the Bible tells me so. Little ones to Him belong, they are weak but He is strong. Yes, Jesus loves me, Yes, Jesus loves me, Yes, Jesus loves me, the Bible tells me so.")

1. Jesus said, "I am the Way,
"Truth and Life, taught you to pray
"No one comes except through Me,
"Follow I will set you free"

Chorus:
Jesus and Father
Jesus and Father
Jesus and Father
One God with Holy Ghost

2. "I forgive your sins," He said
"My body is the broken bread
"My blood the sacrifice for sin"
So that God can let you in

Chorus

3. "Since you've seen Me here with you
"You have seen the Father too
"I will give you life so free
Believe in God, believe in Me"

Chorus

4. "Tell us if you are the Christ,"
Was their question in the strife.
"You are right, I am the Son,
"Sent from God, the Promised One."

Chorus

5. So they hung Him on a cross
Certain this would be His loss
But He rose in victory
Seen by hundreds now set free

Chorus

Chapter Summary

1. Jesus said, "I am the Way, the Truth, and the Life. No one comes to the Father except through Me."

2. He forgave sin. (Only God can forgive sin.)

3. He said, "He who has seen Me has seen the Father" and "I and My Father are one."

4. He responded to the question, "Tell us if you are the Christ!" with "I am."

5. He said, "Before Abraham was, I AM." Jesus claimed to be older than Abraham, and claimed the name of God for Himself, the name "I AM".

6. Jesus was seen in a glorified body (transfigured), conversing with Moses and Elijah.

7. He rose from the dead, with a living body, able to eat, and be touched. Jesus alive after his death on the cross was witnessed by more than 500 people.

Here is a memory aide:

"I AM the Shining Resurrected One Who forgives sin."

- There are 3 "I AM's": I am the way…, Before Abraham was…, Tell us if You are the Christ…
- Resurrected
- Shining: Transfigured
- One: I and My Father are One, He who has seen Me has seen the Father
- forgave sin, and was sinless

C.S. Lewis in his classic <u>Mere Christianity</u> put it this way. "I am trying here to prevent anyone saying the really foolish thing that people often say about Him: "I'm ready to accept Jesus as a great moral teacher, but I don't accept his claim to be God." That is the one thing we must not say. A man who was merely a man and said the sort of things Jesus said would not be a great moral teacher. He would either be a lunatic—on a level with the man who says he is a poached egg—or else he would be the Devil of Hell. You must make your choice. Either this man was, and is, the Son of God; or else a madman or something worse. You can shut Him up for a fool, you can spit at Him and kill Him as a demon, or you can fall at His feet and call Him Lord and God. But let us not come away with any patronizing nonsense about His being a great human teacher. He has not left that open to us. He did not intend to."

CHAPTER 6

HOW CAN YOU ANSWER A MORAL RELATIVIST?

If someone claims to be a moral relativist, with a statement like, "What is right or wrong has to be decided by each individual person for her or himself," what one question can you ask that will defeat this belief? Explain why this question undermines their moral relativism belief.

Here is the question: Is there anyone, anywhere, doing anything you believe is absolutely wrong, something that you believe they absolutely should stop doing no matter what they personally believe about the correctness of their behavior?

ANYONE   ANYWHERE   ANYTHING ABSOLUTELY WRONG?

If you do, and everybody does, that means you do believe that there is some kind of absolute moral standard everybody should abide by regardless of their individual beliefs.

This absolute moral standard is a belief that some things should not be done regardless of how a person feels about them. The Nazis who exterminated Jews didn't feel it was immoral. Terrorists who beheaded journalists believed they were doing service to God. ISIS soldiers who made their female captives into human sex slaves believed they had the right to do this. Nevertheless we believe that what they did was absolutely wrong. And this demonstrates that we humans do believe in absolute moral standards even if we try to convince ourselves that we don't.

For Christians the reason we feel human life and freedom are precious is that we are "made in the image of God." Because God made us morally sensitive and creative we have reason to believe that what we do matters, that we can change things for the better, that freedom to choose (free will) is critically important.

For the non-believer there is no basis for morals and no basis for human rights. If morals are determined by the majority what happens when the majority decides not to support human rights? What happens if the majority decides it is best to exterminate the minority? What if the majority opinion—as it is in many Islamic countries— is that women have less worth than men? Or that slavery is acceptable? Or that sex is nobody's business except the people doing it even if it involves adults having sex with children?

For a more complete description of why moral relativity is not supported by either faith or reason, please see The Reason for God by Timothy Keller.

Chapter Summary

Moral relativists are forced to the conclusion that there are moral absolutes by the question

Is Anyone—Anywhere—doing Anything—you believe is Absolutely Wrong? (Even if they believe what they are doing is right)?

Living in harmony with God includes confronting error. This error— that every person has to decide what is moral for himself or herself—is deadly.   If you allow it to go unchallenged it is assumed that you agree with this statement. Your friends and colleagues need to understand why it is a serious error.

When you hear this statement don't keep quiet. Speak up. Challenge this with the question...Is there anyone, anywhere, doing anything, right now, that you believe they should absolutely stop doing, even if they feel they are doing nothing wrong? Follow up with some questions, like "What if your culture decides women have less worth than men? Many Muslim majority countries share this belief." "What if the majority in your society believes killing female babies is acceptable? Female infanticide is practiced in China and India, where cultural norms value male children over female children." "Should abortion be allowed to terminate pregnancies based upon the predicted sex of the fetus? The selective abortion of female fetuses is most common in China, India, and Pakistan."

Moral absolutes are built into humans by God. It is part of conscience. This is powerful evidence for God. Recognizing that there are moral absolutes is critical to belief in God. Loving Him includes obeying His commandments.

If God is the source of the moral code, if the Bible is God's word on what is moral, we don't have to guess about what is right and wrong. We only have to decide if we are going to obey God or rebel against Him—if we are going to choose our own way or His way. There are only two choices. We are either FOR God or AGAINST Him.

Chapter 7

WHAT IS PASCAL'S WAGER?

THE TRIAL— ILLUSTRATION OF PASCAL'S WAGER

What is Pascal's Wager? What is the Atheists' Wager? What is the difference, and why does it matter?

Pascal's Wager was devised by the seventeenth-century French philosopher, mathematician and physicist Blaise Pascal (1623–62). He applied logic to the question: Is it better to believe in God and live as though God exists, or not?

He set up the wager we all make in this way. All humans bet with their lives either that God exists or that God does not exist. Given the possibility that God does exist and assuming infinite gain (eternity in heaven) or infinite loss (eternity in hell) associated with this choice—a rational person should live as though God exists and seek to believe in God. If God exists this person would go to heaven. If God does not actually exist such a person will have, at most, only a finite loss, perhaps some pleasures or luxuries during life.

If a person lives as though God does not exist and does not seek to believe in God, if God exists this person would suffer infinite loss (eternity in hell). If God does not actually exist such a person would have, at most, some extra pleasures or luxuries during life.

What is the Atheist's Wager? It is similar to Pascal's wager. It agrees that you bet your life! But it argues that you don't really have to believe in God—you only have to live a good life. The Atheist's Wager was formulated by the philosopher Michael Martin and published in his 1989 book Atheism: A Philosophical Justification. It is an atheist's response to Pascal's Wager regarding the existence of God. It has the same premise but reaches a different conclusion.

The Atheists' Wager is stated like this: You bet your life. Therefore a rational person should choose to live a good life. If God actually exists this person would go to heaven. If God does not actually exist this person leaves a good legacy. The atheist says the choice is actually "live a good life" versus "lead an evil life."

The Atheists' Wager includes two logical fallacies including a false premise and a half truth. The false premise is this: If God actually exists, a person who "chooses to live a good life" would go to heaven anyway, even if this person does not believe in God or have faith in God. The half truth is this: A rational person should "choose to live a good life". It sounds right. But the problem is, what does "live a good life" mean? Who defines it? What is a morally good life? Does the individual decide it? Or is it defined by God's laws?

The real question here is "What does God actually require?"

The answer is found in the Bible in Hebrews 11:6 "But without faith it is impossible to please Him, for he who comes to God *must believe that He exists*, and that *He is a rewarder of those who diligently seek Him."*

It is clear that humans have a built in moral compass. This "moral law" is independent of beliefs and is found in all cultures. It supports fairness and courage, justice and honesty, loyalty and kindness. It prohibits selfishness and greed, cowardice and cruelty, cheating and bullying. When you recognize that this moral law is real you recognize the lawgiver too. Morals don't arise by chance. The law of survival of the fittest—natural selection—is opposed to the moral law. The moral law protects the weak and disadvantaged.

It's not only clear that we know the moral law, but it's also clear that every one of us violates this moral law. God calls this sin. We know what the law says to do. We don't do it. This law is absolute. It tells us what we should do, no matter how hard or dangerous it is to do it. This creates a dilemma for us. We know God is the source of the moral law. We agree with Him in our disapproval of human avarice and exploitation. But we disobey the law anyway. And this puts us in opposition to God. We would like to have Him make an exception in our case, but we know He must uphold His law. He wants us to be right with Him. He wants us to be holy. He wants to save us from destruction. But if we break His laws we must face the penalty. When a government makes laws against murder, and the penalty is death, the government wants you to obey the law. It wants to prevent murder. But if you murder, you must expect— and it would be just—for you to be tried, convicted, and executed. That's how the law works. It's there for your good, but the penalty must be paid if you break it.

We are under threat of death because of our sin. We have made ourselves enemies of God. But He has made a way out. If we accept Jesus as Lord, if we put God in command of our lives and give up our right to rule ourselves, He enters our lives, makes us sons and daughters, and gives us eternal life in His presence. He doesn't demand perfect behavior. In time, He will make us perfect. But he does demand obedience. That means that if we go to Him and confess our sins we will be forgiven and restored to His family.

Rick Warren says it this way, "You were made to last forever. Your body will die, but you will go on, in one place or another." Betting that God exists is a good bet. You do bet your life and your life is precious.

Don't wait. Commit your life to the Lord Jesus Christ, and know the peace that passes all understanding.

"The Trial" is a fantasy to illustrate Pascal's Wager. It describes what might happen to Michael Martin, the originator of The Atheists' Wager, at his death, if God is real after all. It is his custody hearing in the Court of Heaven at Judgment. It is a classroom drama.

THE TRIAL

Actors:
- Judge, wearing black robe and wearing crown
- Bailiff, with clip board and papers
- Clerk, with computer
- Archangel Michael, carrying large sword
- Fallen Angel Lucifer, in black tux t-shirt, wearing red curling horns
- Michael Martin, carrying large book labeled "ATHEISM"
- Student 1
- Student 2
- Student—John Misled

[Narrator] The setting is Supreme Court, heaven, Future.

At a table in front of the class, facing the front, sit Lucifer, Michael Martin, and Archangel Michael. Behind them is the gallery [the class]. The witness box is at the right front of the class, to the left of the Judge's bench, facing the gallery. All the other witnesses are sitting in the gallery. Bailiff is sitting to the right of the Judge, behind a table, located perpendicular to the bench. The clerk is sitting next to the bailiff.

Clerk [stands, announces in loud voice]: "All rise!" [The Judge, in long black robe, enters from the left side of the courtroom, walks to the Judge's bench, and sits down.]

Clerk: "You may sit." [All sit down]

Bailiff [Stands, Reads, in loud voice]: "Hear ye. Hear ye. The Honorable Court of Eternal Life is now in session. First case: In the matter of the permanent custody of the soul of Michael L Martin, God of the Universe represented by Archangel Michael, and Fallen Angel Lucifer, Custodian of Hell, Petitioners." [Bailiff sits]

Judge: "We come here today to hear evidence and adjudicate final custody of the soul of Michael L. Martin. Archangel Michael, are you prepared to open?"

Archangel Michael [Stands]: "I am." [He strides to the front of the courtroom on the right, to the left of the Judge, and addresses the gallery and the Judge, facing sideways] "God made the path to eternal life plain both through his written word, the Bible, and through the life of His Son Jesus Christ. God is merciful and fair. Your Honor, we ask you to determine that Mr. Martin has met the qualifications for eternal life in Heaven." [He returns to his desk, facing the Judge, and sits]

Judge: "Fallen Angel Lucifer, are you prepared to state your case?"

Lucifer: [Stands] "I am." [He strides to the right front of the courtroom, and addresses the gallery and Judge, standing sideways.]
"It is plain that this soul should be mine. I would like the court to hear from him directly before he is remanded into my custody for permanent residency in the seven circles of hell. I call my first witness, Michael L Martin."

[Martin comes forward, enters the witness box, carrying his big book.]

Clerk [standing beside the witness box]: "Raise your right hand. Do you swear to tell the truth, the whole truth, and nothing but the truth, so help you God?"

Martin: [Drops his right hand to his side.]   "Well, I'll swear to tell the truth but I certainly won't say 'so help me God' because I don't believe in God. Nobody can really know 'the truth' anyway.   There's no such thing as absolute 'truth' for everybody so" [He raises his right hand] "I swear to tell *my version* of the truth." [Clerk shrugs, turns, and walks away. Martin sits down in the witness box]

Lucifer: [Standing beside the witness box] "Dr. Martin. You held a PhD in philosophy. You were famous. You were persuasive, influential, an intellectual of the highest standing."

Martin: "Well, thank you. Yes I *am*. I got my PhD in Philosophy from Harvard University and *I am* Professor Emeritus at Boston University. *I am* a recognized specialist in the Philosophy of Religion."

Lucifer: "*Were!*" [Lucifer looks at Martin curiously, as if he suspects that Martin doesn't yet know that he is dead, and at the final judgment] "Did you write a book, published in 1989, with the title of Atheism: A Philosophical Justification?"

Martin: "Certainly! It was some of my best thinking. Here it is right here!" [He holds up the book he has been carrying, showing ATHEISM on the spine, and NO GOD on the cover.]   "My other books, of which I am particularly proud, include The Case Against Christianity, The Impossibility of God, and The Cambridge companion to Atheism."

Lucifer: "And did you in your book Atheism, A Philosophical Justification, propose a rebuttal to Pascal's Wager called the Atheist's Wager?"

Martin: "Yes! It's something I'm very proud of. Pascal was an old fashioned thinker. He thought you should believe in God to hedge your bet. I agree that each person is betting his life. But I demonstrated that it was entirely unnecessary to *believe in* God to still get the same result."

Lucifer: "How is that? Please explain."

Martin: "Well, it's only reasonable that the important thing is not that you *believe* in God but that you live a good life instead of an evil life. That way, on the *unbelievably remote* chance that God exists you get into heaven anyway."

Lucifer: "Define the terms, 'live a good life' and 'live an evil life' for me."

Martin: "Well, of course, morality has to be defined by the individual. What is right and wrong is relative. 'Live a good life' means you are true to your ideals of justice and fair play. What is evil is doing things *you feel* are wrong. It's clear that if a god existed at all that he or she would certainly not punish someone for their sincerely held beliefs even if they were, ultimately, wrong."

Lucifer: "So as long as you were acting in accord with your sincerely held beliefs, God, *if he or she existed of course*, would certainly grant you eternal life in heaven."

Martin: "Of course! Now you've got it!"

Lucifer: "Do you understand who the judge is, here in this court?"

[Martin looks at the Judge, does a double take, his face has a look of horror!]
Martin: "OMG!"

Lucifer: "Yes indeed!" [He waits for the agitated Martin to calm down]

Lucifer: "On what basis did you decide that living a good life was what God required to get into heaven?"

Martin: "Why, it's only reasonable. Res ipsa loquitur [pronounced: Ray eep-sa low quee tur]. It speaks for itself."

Lucifer: "Did you consult any sources other than your reason to find out what God requires?"

Martin: "You mean some religious book like the Bible?"

Lucifer: "Yes."

Martin: "That's ridiculous! The Bible is completely untenable, falsified history, unscientific hogwash. The creation story is silly. The Bible is just too incredible. And I mean just what I said. In-credible. Not believable! What *is* credible is that everything started with the Big Bang, and evolution supplied all the rest. The Bible stories are impossible like the one where God stops the sun in the sky. It's naive to believe that Jesus did actual miracles and especially that he was God."

Lucifer: "Did you have access to the Bible?"

Martin: "Of course. I read it many times and used it as an example of creative story telling, poetry, myth, legend, and wishful thinking."

Lucifer: "Did you ever read in the Bible about what God actually requires?"

Martin: "No. No point doing that. It couldn't be true."

Lucifer: "No more questions, your Honor." [He turns away with a wicked grin on his face, sits down.]

Judge: "Any questions from Archangel Michael?"

Michael: [Stands] "No, Your Honor." [Sits back down]

Clerk: "The witness may step down."

[Martin stands, walks back to his place at the attorney's table, sits]

Lucifer: "I call _____ "(actual name of student 1: S1)

[S1 stands, walks to the witness box, stays standing]

Clerk: "Raise your right hand. Do you swear to tell the truth, the whole truth, and nothing but the truth so help you God?"

S1: "Jesus said not to swear by anything, just let your yes be yes and your no be no. I *affirm* that I *will* tell the truth, the whole truth, and nothing but the truth."

[The clerk turns and walks to his desk. S1 sits down.]

Lucifer: "Do you know what God requires for eternal life with Him?"

S1: "Yes. You have to start with belief. In Hebrews 11:6 it says that without faith it is impossible to please God, because anyone who comes to Him must believe that He exists and that He rewards those who earnestly seek Him. So belief that God exists is required."

Lucifer: "What do you mean by the term, 'believe that God exists?'"

S1: "It means to believe that God created the world, and that it didn't just come about by itself. It means that He is the only God, the one true God, the God who revealed himself in the Bible."

Lucifer: "Anything else?"

S1: "It means that because He created the world that it is planned and has a purpose. It is not an accident. It means that he is everywhere present and every time present, hears everything, even our thoughts, and knows everything even before it happens."

Lucifer: "What about belief in Jesus?"

S1: "It means that you believe that God revealed what He is like by sending his Son in the person of Jesus to show us what God is like. It means that Jesus is both God and man and is able to save us from our sins."

Lucifer: "No further questions, Your Honor." [He turns, sits down at the attorney desk]

Judge: "Archangel Michael?"

AM: [Standing] "No, Your Honor." [Sits back down]

Clerk: "The witness may step down." [S1 rises, moves out of the witness box, returns to his seat in the gallery]

Lucifer: [Standing up] "I call _____ " [Student 2, S2, playing self]

[S2 rises, walks to witness box, stays standing]

Clerk [standing by witness box]: "Raise your right hand. Do you swear to tell the truth, the whole truth, and nothing but the truth, so help you God?"

S2: "I do!"

Lucifer: "I'm going to read you a statement from the Bible: It is written: 'He rewards those who earnestly seek Him.' What does that mean?"

S2: "It means that God hears our prayers, even our thoughts, and he answers the prayers of those who believe He is there and who are sincere in their pursuit of relationship with God. He may not answer the way we want, but He hears and answers."

Lucifer: "And what does this mean? 'For God so loved the world that he gave His only begotten Son, that whosoever believes in Him should not perish, but should have eternal life.'"

S2: "It means just what it says. If you believe in God that means you trust Him and accept Jesus to be your Savior. That's what 'believing IN God' means."

Lucifer: "Even I believe God exists."

S2: "True, but you don't TRUST him."

Lucifer [scowling and frowning]: "I'm finished with my questions, Your Honor." [He turns, walks to attorney table, and sits]

Judge: "Archangel Michael. Any further questions of this witness?"

AM: [Stands] "Yes, your Honor."   [He walks up beside the witness box] "Please explain what GRACE means."

S2: "It means that because I have accepted Jesus as my Lord that I have been made a child of God, and that my acceptance in His family and the forgiveness of my sins is a free gift. It's not something I can earn with good behavior."

AM: "So 'living a good life' is not enough?"

S2: "No, it isn't. No person can live a life free of sin. And sin separates us from God. He is holy and nothing unclean can come into His presence. So the only way we can get to God to enjoy the benefits of being his children is to accept the free gift he has given by accepting Jesus Christ as Lord and Savior."

AM: "Nothing else?"

S2: "My hope is built on nothing less than Jesus' blood and righteousness."

AM: [Turns to face the Judge] "I am done with my questions, Your Honor."
[He returns to the table, and sits]

Clerk: [standing] "You may step down." [S2 stands, moves out of witness box, returns to the gallery]

Lucifer: [standing] "I call John Misled"

[John rises from his seat in the gallery, walks to witness box, and stays standing.]

Clerk: [standing by witness box] "Raise your right hand. Do you swear to tell the truth, the whole truth, and nothing but the truth, so help you God?"

John: "I do." [the clerk returns to his seat, John sits down.]

Lucifer [standing next to witness box]: "Were you a student of Professor Martin?"

John: "Yes. I took 'Philosophy of Religion' from Dr. Martin at Boston University. He was one of my favorite teachers."

Lucifer: "Why?"

John: "He told us about reality, how it was impossible for god to exist, how the miracles of the Bible were totally unscientific, how any really enlightened person would certainly come to the conclusion that there *was* no god. He was charismatic, exciting."

Lucifer: "Did you believe in God before you took his course?"

John: "Oh yes. I was from a Christian home and attended Fresno Christian High School before I went away to college. I guess I was pretty naïve. I believed in God and Jesus because that's how I had been raised. But when Professor Martin challenged me to show him some good reasons why my beliefs in God and Jesus were true I didn't have any convincing answers. It was clear that he knew a lot and was an expert in religion so I was pretty sure he was right."

Lucifer: "What did you do then?"

John: "Well, I decided that if there was no God and that morals were relative and if there was really no life after death I better start having a good time, because life was too short to be following a lot of useless rules like the Ten Commandments. So I did it all—sex, drugs, partying, shoplifting—I had a lot of fun at first. But when I died of an overdose I found myself here. Was I ever surprised!"

Lucifer: "No further questions." [Lucifer turns, walks to attorney table, sits down]

Judge: "Archangel Michael?"

AM: [Standing] "No further questions, Your Honor."

Clerk [Stands, to the witness says]: "You may step down." [John stands, leaves the witness box, returns to the gallery]

Lucifer: [He stands, turns, looks up to the Judge] "Your honor, the soul of Michael Martin belongs to me. It is written, 'Whoever causes one of these little ones who believe in Me to stumble, it would be better for him if a large millstone were hung around his neck, and he were thrown into the sea.' I rest my case." [reference Mark 9:42] [Lucifer turns, returns to his seat at the attorney table facing the front of the courtroom.]

Judge: "Archangel Michael, Do you choose to contend for his custody?"

Archangel Michael: [stands up] "I do. I would like to recall the first witness. With your permission, Your Honor, I would like to hear from him."

[Martin rises, returns to the witness box, still carrying his book, sits down. He is trembling now.]

AM: "Dr. Martin, you testified that you tried to live a good life. Where would *you* prefer to be in custody?"

Michael Martin: [He stands up, angry, shouting, holding his book up in the air] "I don't prefer any custody at all. You are trying to force me into a false dichotomy. (pronounced die caught Oh me).
I want to be free. I want to do it MY WAY."
[He storms out of the witness box, looks around, sits down in his seat.]

END

[Teacher, standing facing the gallery—the class]:

Ladies and Gentlemen of the Jury, please discuss how you would decide this case. Elect a foreman. Consider the evidence. The foreman will present the recommendation to the Judge.

First, come to a decision about WHAT you would recommend to the Judge for the outcome of this custody hearing and WHY you would make that recommendation. Put it in writing. If you cannot come to a unanimous decision explain both the majority and the minority points of view.

[When this is complete the teacher asks the next question]

Second, WHERE do you think John Misled is spending eternity? WHY? When you have come to a decision notify me that you are ready to proceed.

Additional discussion questions:
What do you think of Michael Martin's statements:
● "Nobody can know the truth anyway"

- "my version of the truth"
- "morality has to be defined by the individual"
- "what is right and wrong is relative"
- "evil is doing things you feel are wrong"
- "god would not punish you for your sincerely held beliefs even if you were wrong"

2. What about others' statements:
- "no person can live a life free of sin?"
- "my acceptance in His family and the forgiveness of my sins is a free gift. It's not something I can earn with good behavior."
- "when Professor Martin challenged me to show him some good reasons why my beliefs in God and Jesus were true I didn't have any convincing answers." Is this true of you?

3. How does this relate to CS Lewis' view that because humans are immortal (human spirit is immortal) the natural trajectory of rebellion against God is infinite separation from Him (Hell)?

Chapter 8

What do Christians believe will happen to people who die without accepting Jesus Christ as Savior? Will they go to hell?

We simply don't know. It is speculation and not completely answered in the Bible. There are some who believe that all people will be saved and will have a chance to change their minds and believe in God after death. This is called "Universalism." There are some who believe that if you have not accepted Jesus as your savior by the time of your death you will go to hell. There are some who believe "once saved, always saved," that if you have once accepted Jesus as Savior, that no matter what you do or say, no matter what your behavior or future beliefs you will go to heaven. There are some who believe that if you have never heard of Jesus or had a chance to believe in Him that you will be judged by a different standard. There are Bible verses that can be used to support each of these beliefs especially when taken out of context or not compared with other Bible verses.

CS Lewis put it like this. "Is it not frightfully unfair that this new life should be confined to people who have heard of Christ and been able to believe in Him? But the truth is God has not told us what His arrangements about the other people are. We do know that no man can be saved except through Christ; we do not know that only those who know Him can be saved through Him."

We do know that no man can be saved except through Christ; we do not know that only those who know Him can be saved through Him.

C.S. Lewis

A rational Christian belief can be summarized thus: Jesus offers us a way to be right with God. It is not based on our actions but on God's action in bringing Jesus to earth to be our savior. If we accept Him as our Lord (we put ourselves under His command) we will be saved and go to heaven. If we change our minds, that is, after knowing God we turn from him in disbelief we are traitors, joining the enemy, choosing separation from God. That's why "once saved, always saved" is not supported by scripture. Just as you can terminate your natural life by killing yourself you can choose "spiritual suicide." This doesn't mean that when Christians go wrong they are condemned to hell. As long as they still believe, as long as their trust is in Jesus they continue to be forgiven. Again from CS Lewis, "A Christian is not a man who never goes wrong, but a man who is enabled to repent and pick himself up and begin over again after each stumble—because the Christ-life is inside him, repairing him all the time."

Those who spend much time speculating about what happens to all the others—those who have never heard the good news, those who follow a different religion or no religion but who behave according to God's law—would do well to follow Martin Luther's advice: It's a question we should not waste time debating because the Bible doesn't answer it.

Jesus said, "I am the way, the truth, and the life. No one comes to the Father except through Me." This is John 14:6. Jesus is the door to heaven. He makes the decision. It's not for us to speculate about what happens to each individual at death.

Our obligation is to obey His commandments. "Love one another." "Go therefore and make disciples in all the nations, baptizing them in the name of the Father and of the Son and of the Holy Spirit, teaching them to observe all things that I have commanded you." "Forgive one another." "Do not lay up for yourselves treasures on earth, where moth and rust destroy and where thieves break in and steal; but lay up for yourselves treasures in heaven, where neither moth nor rust destroys and where thieves do not break in and steal. For where your treasure is, there your heart will be also." "Judge not, that you be not judged. For with what judgment you judge, you will be judged; and with the measure you use, it will be measured back to you." "Therefore, whatever you want men to do to you, do also to them, for this is the Law and the Prophets." "For God so loved the world that He gave His only begotten Son, that whoever believes in Him should not perish but have everlasting life." If we do this, Jesus said, "By this all will know that you are My disciples, if you have love for one another."

CHAPTER 9

WHAT ABOUT SEX?

Christians are torn by the conflict between cultural norms and Christian ethics. We are bombarded from all sides by the idea that sex is about personal satisfaction and freedom. Hookups and friends with benefits are euphemisms for casual sex, sex with no commitment expected or implied, sex as a commodity instead of a relationship. The women's liberation movement in its promotion of sex with no strings attached produced exactly the opposite result from that which it intended. It empowered men and devalued women as men used women as objects to satisfy their desires, to be discarded or ignored as soon as they found a better value elsewhere (younger, prettier, less demanding, more willing to have sex anywhere and anytime). This kind of selfish sex feeds the masturbation industry (pornography) as well. It's all about me. It's all about what I can get, not what I can give.

But sex was never intended to be casual or selfish. It is demeaning to both men and women to be treated as objects instead of persons. God's view of sex is that it is good. He invented sex to be enjoyed in a long term committed relationship. He set the rules for how to enjoy it based on what He knows is best for us. It's like the owners' manual for your car. It tells you how to maintain it for best results. It will still run if you abuse it. But it will also do long term damage and shorten its useful life if you fail to follow the recommendations. Sex is like that.

First you need to understand what God says about sex and apply it to your own life. Second you must not be judgmental about what others are doing because Jesus was clear: "Do not judge others, and you will not be judged." This is from Luke 6:37. Christian sexual morality is called "chastity." As put by C.S. Lewis, "Chastity is the most unpopular of the Christian virtues. There is no getting away from it; the Christian rule is,

'Either marriage, with complete faithfulness to your partner, or else total abstinence.'"

Chastity is the must unpopular of the Christian virtues…the Christian rule is, either marriage, with complete faithfulness to your partner, or else total abstinence.

C.S. Lewis

There is a little Sunday School song, "Everything that God Made is Good." It goes, "Who made the birds, God did! Who made the birds, God did! Who made the birds, God did! And everything that God made is good!"

This goes for sex too. In the beginning we are told in Genesis 1:27-28, "So God created man in His own image, in the image of God He created him, male and female He created them. Then God blessed them, and God said to them, 'Be fruitful and multiply: fill the earth and subdue it.'" The first command, before the fall, before there was any sin, was a sexual command to the first male and female humans He created. "Then God saw everything that He had made, and indeed it was very good." Genesis 1:31.

This "good sex" was related to marriage from the very beginning. In Genesis 2:24 it says, "That is why a man leaves his father and mother and is united to his wife, and they become one flesh." Jesus reiterated this in Mark 10: 2-9. Some Pharisees came and tested Him by asking, "Is it lawful for a man to divorce his wife?" "What did Moses command you?" He replied. They said, "Moses permitted a man to write a certificate of divorce and send her away." "It was because your hearts were hard that Moses wrote you this law," Jesus replied. "But at the beginning of creation God 'made them male and female.' 'For this reason a man will leave his father and mother and be united to his wife, and the two will become one flesh.' So they are no longer two, but one flesh. Therefore what God has joined together, let no one separate." When they were in the house again, the disciples asked Jesus about this. He answered, "Anyone who divorces his wife and marries another woman commits adultery against her. And if she divorces her husband and marries another man, she commits adultery."

These are hard words. Jesus confirmed they were true. Sex, like everything in God's economy, comes with conditions on how to use it. God not only created sex, He encourages it in marriage. Lawful sexual acts are those permitted in marriage between a woman and man bound in holy matrimony. This is specified in the sixth commandment, "You shall not commit adultery," Exodus 20:14. All other sexual acts are offenses against God's ordinances and therefore sinful. This includes sexual intercourse between unmarried adults, sex between a married person and someone not their spouse, masturbation (sex with yourself), sex between men and men, sex between women and women, sex with children, and sex with animals.

Jesus made it clear that sexual immorality includes even impure thoughts. God knows our every thought. Absolutely nothing is hidden from Him. In Matthew 5:27 Jesus said, "You have heard that it was said, 'You shall not commit adultery.' But I tell you that anyone who looks at a woman lustfully has already committed adultery with her in his heart." It is clear that we are all guilty before God. No one has passed this test except Jesus Himself. Is this cause for despair, for giving up trying to be right with God? No, for He made a way out. As long as we want to be right with God—as long as we are willing to return and admit our guilt and submit our wills to Him—He will, like the father of the prodigal son, welcome us back to His family with open arms.

The Bible is clear about sex.

Leviticus 18:22-25. "You shall not lie with a male as with a woman. It is an abomination. Nor shall you mate with any animal, to defile yourself with it. Nor shall any woman stand before an animal to mate with it. It is perversion. Do not defile yourselves with any of these things, for by all these the nations are defiled, which I am casting out before you."

Leviticus 20:13. "If a man lies with a male as he lies with a woman, both of them have committed an abomination."

Leviticus 20:15-16. "If a man mates with an animal, he shall surely be put to death, and you shall kill the animal. If a woman approaches any animal and mates with it, you shall kill the woman and the animal.

Leviticus 20:23. "And you shall not walk in the statutes of the nation which I am casting out before you; for they commit all these things, and therefore I abhor them."

Mark 7:20-23. "What comes out of you is what defiles you. For from within, out of your hearts, come evil thoughts, sexual immorality, theft, murder, adultery, greed, malice, deceit, lewdness, envy, slander, arrogance and folly. All these evils come from inside and defile you."

I Corinthians 6:9-11. "Do you not know that the wicked will not inherit the kingdom of God? Do not be deceived: Neither the sexually immoral nor idolaters nor adulterers nor male prostitutes (boys kept for sexual relations with a man) nor homosexual offenders nor thieves nor the greedy nor drunkards nor slanderers nor swindlers will inherit the kingdom of God. And that is what some of you were. But you were washed, you were sanctified, you were justified in the name of the Lord Jesus Christ and by the Spirit of our God."

Romans 1:25-32. "They exchanged the truth of God for a lie, and worshiped and served created things rather than the Creator—Who is forever praised. Amen. Because of this, God gave them over to shameful lusts. Even their women exchanged natural relations for unnatural ones. In the same way the men also abandoned natural relations with women and were inflamed with lust for one another. Men committed indecent acts with other men, and received in themselves the due penalty for their perversion. And even as they did not like to retain God in their knowledge, God gave them over to a debased mind, to do those things which are not fitting; being filled with all unrighteousness, sexual immorality, wickedness, covetousness, maliciousness, full of envy, murder, strife, deceit, evil-mindedness, whisperers, backbiters, haters of God, violent, proud, boasters, inventors of evil things, disobedient to parents, undiscerning, untrustworthy, unloving, unforgiving, unmerciful; who, knowing the righteous judgment of God, that those who practice such things are deserving of death, not only do the same but also approve of those who practice them."

Hebrews 13:4. "Marriage should be honored by all, and the marriage bed kept pure, for God will judge the adulterer and all the sexually immoral."

Jude 7. "In a similar way, Sodom and Gomorrah and the surrounding towns gave themselves up to sexual immorality and perversion. They serve as an example of those who suffer the punishment of eternal fire."

The Bible makes it plain what God approves— sex between a married husband and wife—and what He does not approve—anything else. This is made crystal clear in Ephesians 5:3-4. "Among you there must not be even a *hint* of sexual immorality, or any kind of impurity, or of greed, because these are improper for God's holy people. Nor should there be obscenity, foolish talk or coarse joking, which are out of place."

This is difficult to accomplish especially in a culture that glorifies sex in clothing, movies, magazines, books, billboards, music, advertising and sports. We are assaulted with sexual images constantly. It seems impossible to avoid "even a hint of sexual immorality." But it is possible. The Bible makes it clear that for every temptation God provides a way out. And if we ask for help and don't plead that we are helpless to obey He will make it possible to live a pure life. And when we fail, for we surely will, He is still our loving Father ready to forgive when we are ready to try to obey Him again.

None of this means that Christians are free to hate or fear those who violate God's sexual laws. Jesus commanded his followers to love and pray for everyone. But this love does not imply approval of their behavior. Jesus was criticized for eating and drinking with sinners. He said, "Those who are well have no need of a physician, but those who are sick. I did not come to call the righteous, but sinners, to repentance." This is Mark 2:17. He associated with sinners even while confronting them with their sins and the need to repent.

There are many men and women who have homosexual tendencies, or who do not identify with their biologic gender, or who have sexual attractions to both genders. They make up a significant number in the human population. The cause of these tendencies is not clear. But acting on these tendencies is contrary to God's laws. It is sin. The only lawful response for these individuals in God's economy is chastity, making the decision to forego acting on these tendencies. But whether they are acting on these tendencies or not Christians must accept them with compassion and respect, avoiding discrimination against them.

The tendency, the inclination, the desire to have a homosexual relationship is not sin. The desire to refuse your biological sexual identity is not sin. The urge to have sexual intercourse before you are married or outside your marriage is not sin. But it is sin to carry out these sexual actions in defiance of God's ordinances. Chastity is not just desirable but possible. I Corinthians 10:13 says, "God keeps his promises, and he will not allow you to be tested beyond your power to remain firm; at the time you are put to the test, he will give you the strength to endure it, and so provide you with a way out."

There are two errors that Christians fall into with reference to sexual sins. The first is to reinterpret the Bible to say that homosexual, bisexual, transgender and extramarital sexual relations are not sin at all. Some religious leaders have taken the position that homosexual acts are really not sinful, especially if done in a committed relationship, like a marriage between a woman and woman, or a man and man. Some protestant denominations such as the United Church of Christ, Presbyterian Church (USA) and Episcopal Church have accepted this idea and ordain practicing homosexuals in committed relationships, arguing that God certainly couldn't have called loving relationships between committed individuals "sin." These same denominational theologians presume to speak with authority and wisdom about what God really thinks, or what the Bible actually says. But, in fact, they do not have any secret knowledge hidden from the rest of us. The Bible is clear.

The second error that Christians have fallen into is to condemn those involved in homosexual, bisexual, transgender and extramarital sexual relations and to state that those who do such things are "going to hell." But as Tim Keller, pastor of Redeemer Presbyterian Church in New York City said, "being a homosexual doesn't make you go to hell any more than being a heterosexual makes you go to heaven. I happen to know this."

What does make you go to hell? It's not sin. "Everyone has sinned, and fallen short of the glory of God." This is Romans 3:23. So what does cause you to go to hell? It is refusal to accept God's will, refusal to accept His free gift of grace with forgiveness of sin that comes from believing in Jesus. Everyone sins. Those who want to serve God—those who want to love God and obey Him—are accepted in God's kingdom (heaven) by admitting they are wrong (repenting), asking for forgiveness, and accepting God's gift of life.

But those who refuse to submit to God's will—those who do not want to live in God's kingdom (heaven) under His rule, who prefer to do their own will rather than God's will—choose hell for themselves. This applies equally to the self-righteous who vehemently judge homosexuals but who do not heed Christ's command to "love your neighbor as you love yourself." If these Pharisees believe their righteousness (law keeping) will get them into heaven they are committing a grave error. They are in danger of hell for judging others, failing to forgive and failing to love. They need to fall to their knees, admit their lack of love, repent and ask for forgiveness. God's kingdom admits no hate and selfishness.

So love and accept those who have homosexual, transsexual, or bisexual tendencies. Love those who have had an abortion. Love the promiscuous. Avoid every trace of unjust discrimination toward them. They, like everyone else, are called to fulfill God's will in their lives. If they are Christians they are called to chastity. If they are not it is your duty to pray for them. Pray for them and encourage them to conform their lives to Christ.

What should be the Christian's duty about sex? Obey God's commands. Keep sexual activity where it belongs in marriage. Keep yourself virgin until married. Don't divorce because God sees married couples as "one flesh." It is clear from longitudinal social studies that couples considering divorce are much happier five years later if they go to counseling and keep their marriages together than those who give up and divorce. This is what Christians should expect given God's view of holy marriage. It is also clear from these same studies that couples who live together before marriage are far more likely to break up (80% break up before they are married or just after they are married), and far more likely to divorce (only 12% are still married after 10 years) than couples who wait to live together until after they are married.

Couples who live together before marriage are far more likely to break up and far more likely to divorce than couples who wait to live together until after they are married.

The reason for these statistics is clear. Sexual activity outside marriage treats sex as consumer goods and services, or as recreational goods and services. Treating sex as a commodity devalues it so that staying with a person depends entirely on whether sex with that person is seen as a "good value" or not. And when it is not and sex with someone else looks like a better value that person "walks" out of the relationship. Many young Christians settle for cohabitation because they are worried about divorce and they are afraid to get married. It might not work out. It seems much easier to break up if you just live together. In the tension between following Jesus and the overwhelming desire for sexual satisfaction many Christians drop out because it seems incongruous to attend church when you are living together and know you are breaking God's commandments. Cohabitation is not rebellion. It is what the in crowd applauds. It is the cultural norm. Deciding to get married is the bold thing, the courageous thing to do. Isn't that amazing? It is the very thing God asks you to do before you live together.

Sex works best in marriage because the covenant relationship—the promise you make before God, your family and friends—removes sex from the "recreational goods and services" category and moves it to the "loving relationship" category. You promise to remain married "for better or worse, for richer or poorer, in sickness and in health, forsaking all others, until death do us part." That kind of commitment leads to genuine enjoyment of sex, sex with full abandon, knowing your partner is there to stay! This does not mean that sex is always wonderful inside marriage. Like everything else about a relationship sex is often hard work to make it work. Real joy in sex is keeping it where God says it belongs inside a committed marriage.

And when we fail—and all of us will because even looking at someone you're not married to and thinking about sex with them is sin—what He asks is that we admit our fault, come back to Him and beg forgiveness, and try again. He is faithful and just to forgive us our sins and cleanse us from all unrighteousness.

Finally, in everything don't forget what Jesus said when asked which was the most important commandment in the Law. It is recorded in Matthew 22:36-40. Jesus replied, "'Love the Lord your God with all your heart and with all your soul and with all your mind.' This is the first and greatest commandment. And the second is like it: 'Love your neighbor as yourself.' All the Law and the Prophets hang on these two commandments."

Jesus meant exactly what He said. You are called to love your homosexual neighbor, your Islamic neighbor, your atheist neighbor, your neighbor who has had an abortion, your neighbor living with his girlfriend, even yourself when you have been living in sin. This leaves absolutely no room for hatred, avoidance, or ridicule. We are all in the same boat!

CHAPTER 10

WHAT ABOUT ABORTION?

Is abortion murder? What about assisted suicide for those with severe disabilities, or those terminally ill?

Since abortion was legalized by the Roe vs Wade decision in 1973 there have been over 40 million abortions in the United States. This fact doesn't startle us because it is just another statistic. As Joseph Stalin famously said, "The death of one man is a tragedy, the death of millions is a statistic."

In the United States of America case of Roe vs. Wade in 1973 the Supreme Court decided by majority opinion that abortion was a fundamental right under the United States Constitution. Justice Blackmun's majority opinion explicitly rejected a fetal "right to life" argument. It also stated that "*a fetus is not a person* within the meaning of the Fourteenth Amendment."

"a fetus is not a person"
Justice Blackmun's majority opinion 1973

Please note that the 14th Amendment to the United States Constitution was adopted to make black former slaves "persons" with equal rights to other United States citizens. Prior to this time black slaves were not "citizens" and were not included as "men" as stated in the Declaration of Independence. "We hold these truths to be self-evident: that all men are created equal; that they are endowed by their Creator with certain unalienable rights; that among them is life, liberty, and the pursuit of happiness..."

In the Dred Scott decision of 1857 before the 14th Amendment was proposed the Supreme Court of the United States had decided, "In the opinion of the court, the legislation and histories of the times, and the language used in the Declaration of Independence, show, that neither the class of persons who had been imported as slaves, nor their descendants, whether they had become free or not, were then acknowledged as a part of the people, not intended to be included in the general words used in that memorable instrument."   In other words, this Dred Scott Supreme Court decision ruled that slaves, former slaves, and descendants of slaves were not "people," "men," or "citizens."

Slaves, former slaves, and descendants of slaves are not people, men, or citizens
Dred Scott Supreme Court decision 1857

In response to this clearly discriminatory decision the 14th Amendment to the United States Constitution was adopted in 1868 making it clear that "All persons born or naturalized in the United States, and subject to the jurisdiction thereof, are citizens of the United States and of the State wherein they reside. No State shall make or enforce any law which shall abridge the privileges or immunities of citizens of the United States; nor shall any State deprive any person of life, liberty, or property, without due process of law; nor deny to any person within its jurisdiction the equal protection of the laws."

To correct the decision of the Supreme Court in the Dred Scott case required an amendment to the United States Constitution.

Roe vs Wade is a similar case. The Supreme Court majority ruled in 1973 that "a fetus is not a person." Dissenting opinions to Roe vs Wade were written by Justices White and Rehnquist. White wrote, "I find nothing in the language or history of the Constitution to support the Court's judgment. The Court simply fashions and announces a new constitutional right for pregnant women and, with scarcely any reason or authority for its action, invests that right with sufficient substance to override most existing state abortion statutes." He wrote that the Court "values the convenience of the pregnant mother more than the continued existence and development of the life or potential life that she carries." In addition, he criticized the ruling for creating "a constitutional barrier to state efforts to protect human life and by investing mothers and doctors with the constitutionally protected right to exterminate it."

An interesting opponent of Roe vs Wade is Roe herself. In 1995, Norma L. McCorvey (Roe) revealed that she had become anti-abortion and a vocal opponent of abortion. "It was my pseudonym, Jane Roe, which had been used to create the 'right' to abortion out of legal thin air. But Sarah Weddington and Linda Coffee [her attorneys] never told me that what I was signing would allow women to come up to me 15, 20 years later and say, 'Thank you for allowing me to have my five or six abortions. Without you, it wouldn't have been possible.' Sarah [her attorney] never mentioned women using abortions as a form of birth control. We talked about truly desperate and needy women, not women already wearing maternity clothes."

Presidential opinions have been mixed along major party lines. Opponents of legal abortion included Gerald Ford, Ronald Reagan, George H.W. Bush, and George Bush, all Republicans. Supporters of legal abortion included Jimmy Carter, Bill Clinton, and Barack Obama, all Democrats.

A completely different view was present in the Republic of Ireland until 2018. Its constitution reflected Christian values. It provided that "the State acknowledges the right to life of the unborn and, with due regard to the equal right to life of the mother, guarantees in its laws to respect, and, as far as practicable, by its laws to defend and vindicate that right." Ireland's laws allowed abortion only when continuation of pregnancy would put a woman's life (not merely health or other interests) at risk. In 2018 the Irish voters overwhelmingly approved a change to their constitution to allow abortion on demand. It is now legal for a mother to abort her unborn child in Ireland as well as the United States. But God has always had an absolute prohibition against the murder of an unborn human child.

Reasons used to justify abortion include rape and risk to the mother. But less than 1% of all abortions are performed to save the life of the mother. And less than 1% of abortions are performed for rape or incest.

Over 40 million abortions in the United States since 1973.
- ## Less than 1% of all abortions are performed to save the life of the mother
- ## Less than 1% of abortions are performed for rape or incest

Abortion is often characterized as "a woman's right to choose." This is the origin of the "Pro Choice" label, meaning women have the right to choose what to do with their own bodies including abortion. The argument takes many forms. "The fetus isn't really human yet, because it can't survive without its mother." "Women should have the right to decide to have a baby or not." "Abortion is a higher good because the aborted babies are unwanted. They would grow up in squalor and be abused, and many would become victims of human trafficking." "Because the aborted babies die before they can be held accountable, they will go to heaven, and they probably wouldn't if they were allowed to be born. So it's better to allow them to be aborted." "It's my body! I can do what I want with it! It's none of your business what I do with it!"

But the Christian view is the opposite. Human life is sacred because God created humans "in His image." Christians base their objections to abortion on the commandment, "Thou shalt not murder." This is Exodus 20:13. "Do not slay the innocent and the righteous," Exodus 23:7. It is clear from the context of the Bible that this does not refer to warfare, self-defense, or police forces. But abortion is different. Abortion kills an innocent human being. It is murder.

In addition, Christians believe that it is wrong for a mother to make a decision to abort her fetus because her body and the body of her unborn child do *not* belong to her. It is not just "her body." God created it—it is His. He has defined what you are allowed to do with it. And the baby growing inside the womb from the moment of conception is a separate human being, not "tissue," not "your own body." This is not just opinion; it is scientific fact. The fetus has full human potential right from the moment the sperm fuses with the egg and the first human cell is formed that will become a baby.

Even after birth a baby is dependent on the mother for survival. Without food, shelter, and protection the baby would not survive. There is no moral difference between infanticide, murder, and abortion. They are all deliberate premeditated acts to end the life of another human being. Abortion, like murder and infanticide, is a serious sin. The fate of the aborted human fetus is in God's hands. God saves the innocents.

But cooperation in an abortion is a grave offense. It is a crime against human life.

Does this mean Christians believe abortion should be illegal? The Catechism of the Catholic Church says it best. "The inalienable right to life of every innocent human individual is a constitutive element of a civil society and its legislation. The inalienable rights of the person must be recognized and respected by civil society and the political authority. These human rights depend neither on single individuals nor on parents; nor do they represent a concession made by society and the state; they belong to human nature and are inherent in the person by virtue of the creative act from which the person took his origin. Among such fundamental rights one should mention in this regard every human being's right to life and physical integrity from the moment of conception until death." "The moment a positive law deprives a category of human beings of the protection which civil legislation ought to accord them, the state is denying the equality of all before the law. When the state does not place its power at the service of the rights of each citizen, and in particular of the more vulnerable, the very foundations of a state based on law are undermined."

Are you saying you believe that abortions should be against the law?

Yes. I'm saying that abortions should be illegal.

Don't you know that you can't legislate morality?

If you mean you can't produce moral behavior with laws, I agree with you. But we legislate morality all the time. We have legal prohibitions against murder, theft, perjury (false witness), assault, child pornography, prostitution, incest, speeding, running red lights, and cheating on your income taxes. All these things are illegal only because they are immoral. If there is no God, if evolution is all there is then there are no reasons for such moral prohibitions. We don't find it immoral for animals to kill other animals, even animals of their own species. We find it immoral only for humans. We don't find other animals guilty of theft, only humans. And if abiogenesis and evolution is all there is, if naturalism is right that there is no god, then there is no such thing as "free will" either. But we do believe in free will. That's why we legislate against the abuse of free will to prohibit murder, theft, and red light running.

These same considerations apply to other kinds of termination of human life such as euthanasia and suicide. Participating in or assisting someone to put an end to the life of a handicapped, sick, or dying person is morally unacceptable. It is wrong. On the other hand, discontinuing extraordinary medical procedures that are disproportionate to the expected outcome, that are overly ambitious in trying to prolong life, are legitimate. In this case one does not will to cause death. You simply accept your inability to prevent death.

Is there forgiveness for the sins of abortion, euthanasia, and murder? Of course. All sin except for blasphemy of the Holy Spirit is forgivable. What does such forgiveness require? It requires what John the Baptist and Jesus taught: Repent! It requires a turn toward God, a decision that you choose to believe that what He says is right, turning away from evil, making a 180 degree turn to obey God. It means declaring that you are guilty, confessing your sin and asking forgiveness. The Bible is clear. If we confess our sins He is faithful and just to forgive us our sins, and cleanse us from all unrighteousness. This is I John 1:9.

Don't forget that Christians are first of all citizens of the kingdom of God and temporary residents here on earth. We are called to love our neighbors as ourselves. This includes our neighbor who has had an abortion, our neighbor who works at an abortion clinic, and our neighbor who believes it is right to support legal abortion. We can and should pray for them and for our nation. Then, like the example of Brother Lawrence, we should continue to love them and rest in the knowledge that God can correct the situation when He wills.

CHAPTER 11

WHY DO CHRISTIANS SUFFER?

Why do Christians have the same troubles, hurts, and temptations that the world has? Didn't Jesus promise us abundant life?

Jesus gave us a clear picture of what life was like in the parable of "the wise man who built his house upon a rock," from the gospel of Matthew, chapters 5 to 7.

First He gave instructions on how to live a godly life. Be a good influence. Don't be angry without good cause. Don't even think about participating in immoral sexual behavior. Don't swear to tell the truth, just tell it. Answer "yes" or "no" and mean it. Don't sue or try to get revenge when you are wronged (and you will be wronged)—it's better to give in and pay, even if it's unfair. Give and lend to those in need. Love your enemies. Bless and do good to those who hate you and curse you. Pray for those who use you unfairly. Be just like your heavenly Father who sends His rain and sun on believers and sinners alike. Treat others exactly like you would like to be treated. Give quietly. Pray in private. Fast in secret. Forgive, so God can forgive you. He won't forgive you if you don't forgive others. Don't be greedy and hoard money. Love God, not money. If you do this, you won't worry. Worry doesn't help because it doesn't change anything. Judge yourself, not others. Ask your Father for good things and keep on asking. He always gives you what you need. Watch out for false teachers. You can tell who is a false teacher by what their life is like—how they behave in private.

In the same teaching Jesus told us what life is like. Life is like a storm—a hurricane. In each and every person's life there will be devastating trauma that you can't avoid. He gave us a storm warning: expect thunder, lightning, lashing rain, massive flooding, and furious wind. He told us how to prepare. Do God's will. Don't just understand it. Do it! Make God's word the rule for your life. If you do this you will withstand all the troubles, hurts, and temptations of life. Your life will be like a house built on a solid granite hill with the foundation firmly drilled into the rock, the walls built of reinforced concrete, the roof constructed with steel beams, the windows shuttered tightly against the wind. Then when the storms of life come as they certainly will to everyone your house will stand firm. It will be a testament to everyone who sees it that your life was built on obedience to Christ.

But if you know the truth and you don't do it, if you don't obey God's word, your life will be like a house built on sand. It has no foundation. This house is unprepared for storms. When the storms hit, the rain slams down, the flood weakens the footing, the wind shatters the windows, the siding blows off, the roof caves in and the house falls with a crash. Your life will be ruined by the storm.

Both houses are attacked by storms but one will stand firm and one will collapse! What is the difference? Obedience. God is more concerned with your obedience than He is with your looks, grades, career, success, or prosperity. Abundant life has nothing to do with making more money, driving a faster car, living in a bigger house, winning in sports or academics, having more girlfriends or boyfriends, or having the most beautiful face or body. Abundant life is about developing godly character. Your car and your house don't get to go to heaven. But your character does.

Abundant life happens when you love God with all your heart, with all your soul, and with all your strength; and when you love your neighbor as yourself. As Brother Lawrence said, "Our only business in life is to please God." God does not promise that life will be easy for those who put their faith in Him. But He does promise peace that passes all understanding. He promises you a place in His family. He promises He will never leave you or forsake you.

Problems have a purpose. We are told in Romans 8:28 that "Everything that happens fits into a pattern for good for those who love God, who are called according to His plan." Note that this promise is for those who love God. For Christians, God uses problems to develop our character.

Trials, troubles, and problems cause us discomfort. We wish we could avoid illness, death, accidents, lack of money, family disputes, conflict at school or work, failure of others to follow through on commitments and promises. But don't forget that nothing can happen to you that God does not allow. He doesn't cause everything to happen but He does allow it to happen to you. He allows the storms of life. And problems have a purpose. They teach us to trust God.

If everything that happens fits into a pattern for good for those who love God, and if you love God, then every trouble, every trial, every sickness, every accident, every conflict is an opportunity for Him to develop your character, to teach you to trust Him, to teach you patience, kindness, goodness, gentleness, and self-control. As Brother Lawrence said, "seek from Him the strength to endure as much, and as long, as He shall judge to be necessary for you." If you see trials, troubles, and problems as tests God allows for you, you will see them as a favor from God, not as a cause for grief and distress. Don't ask, "Why me, God?" Ask instead, "What do You want me to learn?"

We are often hurt by others. We are offended by the way we are treated, the way we are talked about, the way we interpret comments. Both children and adults intentionally hurt others by their negative words. Rick Warren calls these things "relational nuclear weapons, including condemning, belittling, comparing, labeling, insulting, condescending, and being sarcastic." So why do Christians have to put up with these things?

Hurts are opportunities to teach us to forgive. We are told in the Lord's Prayer, "Forgive us our trespasses as we forgive those who trespass against us." If you think deeply about what Christ has forgiven you, you will find it much easier to forgive someone who has offended you. Accept these opportunities. Forgive! It will build humility, kindness, gentleness, and self-control.

Even temptations are things God allows. God does not tempt. Satan is the tempter, trying to lead us to disobey God. Why does God allow this? Temptations are opportunities to deny ourselves and obey Him. The Bible is clear. With every temptation God makes a way out if we decide to obey Him. He says, "Call on Me in time of trouble." This is Psalm 50:15. First Corinthians 10:13 says, "No temptation has overtaken you except such as is common to man; but God is faithful, who will not allow you to be tempted beyond what you are able, but with the temptation will also make the way of escape, that you may be able to bear it." Temptations are opportunities to obey. The fruit of obedience is joy. Obedience to God makes your life like a house built on rock.

Rick Warren's book, The Purpose Driven Life, details why Christians cannot expect "abundant life" to mean a trouble-free life with material abundance.

To help remember these "fruits" of troubles, trespasses, and temptations, you can sing this little song usually used by the in-crowd to encourage someone to give in to temptation and go along with them. You will probably recognize it. "Every party needs a pooper, that's why we invited you, Party Pooper! Party Pooper!" The melody was written in 1916 as the song, "Pretty Baby." You can find it on YouTube, sung by Dean Martin.

When you are tempted to go along with the crowd, as they sing "Party Pooper" to you, just remember this version.

Every problem has a purpose,
If we love God that is true.
Problem, Purpose! Problem, Purpose!
He lets trials so we'll trust him
He lets hurts so we'll forgive.
Problem, Purpose! Problem, Purpose!

Even when we're tempted
If we choose to obey,
God will make a way out every time.
Every problem has a purpose,
If we love God that is true.
Problem, Purpose! He loves you!

CHAPTER SUMMARY

Jesus promised His followers abundant life. This does not mean a comfortable affluent life. God allows Christians to have the same storms of life as everyone else for a purpose—to develop their characters. Character is more important than accomplishments, career, wealth, and health. Character lasts for eternity. God allows trials so we can learn to trust Him. He lets us be hurt so we can learn to forgive. And He lets us be tempted so we can learn to obey God. As we obey Him, as we do what He asks we gain the fruits of the spirit—love, joy, peace, patience, kindness, goodness, faithfulness, gentleness, and self-control. These are eternal. These are what abundant life is all about.

CHAPTER 12

DO YOU BELIEVE GOD CREATED THE UNIVERSE IN SIX DAYS?

Doesn't the Bible say that the heavens and the earth were created in six 24 hour days about six thousand years ago? How can Christians reconcile this account with scientific evidence that the universe is about 14 billion years old and the earth about 4.5 billion years old?

Genesis 1:1-2 tells us "In the beginning, God created the heavens and the earth. Now the earth was formless and empty, and darkness was over the surface of the deep, and the Spirit of God was hovering over the surface of the waters." These initial two verses of the Bible say nothing at all about "days" and already He had "created the heavens and the earth."

The Bible Old Testament was written in Hebrew, and the Hebrew word for "day" is "yom." Yom is used in several different ways in the Bible. One way is to mean one literal 24 hour day. Another is the way it is used in Genesis 2:4. "These are the generations of the heavens and the earth when they were created, in the day (yom) that the LORD God made the earth and the heavens." It is clear that this meaning includes the entire time of creation, not just one 24 hour day. And in Hosea 6:2 referring to two "days" and third "day" the "yom" is clearly not referring to a single 24 hour day.

One more Bible reference is in order. 2 Peter 3:8 says "But do not forget this one thing, dear friend: With the Lord a day is like a thousand years, and a thousand years are like a day."

Some Christians believe in a "young earth" with the entire creation completed in six 24 hour days. But others who believe in a literal historical interpretation of the Bible find that the Bible account of creation is compatible with scientific findings indicating the age of the universe at 13.73 billion years plus or minus 0.12 billion years, the age of the sun about 4.6 billion years, the age of the earth 4.5 billion years, and the origin of life about 3.5 billion years ago.

Don't let the "young earth" argument divert you from the primary question. It is this: Is God real? Did God create the heavens and the earth or did this just happen by accidental natural means? Did God create life or did life start itself by chance? Did God create species including man or are we all just the result of blind mutation and natural selection?

The answer to all these questions is obvious by now. God is real. He created the heavens and the earth. They did not create themselves. God created life. It requires an intelligent designer to have even the simplest possible living cell. And mutation with natural selection proves to be an inadequate explanation for development of species.

For a detailed explanation about how science and the Bible do not conflict about creation, please see John C. Lennox' book, Seven Days That Divide The World.

Science was invented by God. He likes it. He wants us to know how things work, and how they came to be. The deeper you go in your search for explanations of origins, the more clearly you will see the hand of God.

CHAPTER 13

What about tithes and offerings? How much should Christians give to their church?

When you pledge your life to Jesus Christ you begin to recognize that everything belongs to God. He created it all.   Like the old chorus says, "He owns the cattle on a thousand hills, the wealth in every mine, He owns the rivers and the rocks and rills, the sun and the stars that shine, wonderful treasures more than tongue can tell, He is my Father so they're mine as well, He owns the cattle on a thousand Hills, I know that He will care for me."

The tithe means tenth. God asked that we give the first 10% of our income to Him.   This is a principle clearly described in the Bible even before the 10 Commandments were given to Moses. If we give Him the tithe, He gives the other 90% to us to use as we want, with His blessings. But if we try to keep the whole 100% for ourselves the income will not bless us, but curse us. See Malachi 3:8-12. When we tithe God promises to provide all we need. This is not a "name it and claim it" belief nor is it a formula to become wealthy.   Abundant life is not about having more money and possessions.   But it is about having our needs met, being blessed, as we honor our Father. Offerings are gifts to God in addition to the tithe.

For a detailed explanation please read <u>The Blessed Life</u>, by Robert Morris.

CHAPTER 14

IS IT HARD TO BECOME A CHRISTIAN?

Yes. Because loving someone means giving up your independence. If you love someone you decide to do what pleases them rather than pleasing yourself. You compromise with them. You defend them. You make promises to them. You sacrifice for them. Why? Because loving someone and being loved are what makes life worth living.

We are raised to value autonomy—our freedom to choose anything we want. We are selfish. We are proud of our independence. We want to be in charge. Of course we are always constrained to some degree by our conscience. But conscience can be seared or silenced by hubris. Autonomy is lonely. Pride is competitive. Love is the polar opposite of independence and pride.

As you come to understand that God is the creator of the universe and the only source of life, that He is good and that your soul is immortal, you begin to understand your purpose. You were created to love Him as He loves you with your whole heart. Apart from Him is nothing. Apart from Him is hell.

Being a Christian is both hard and easy. It's easy because God loves you. He wants to make you His child. It's safe there in His home. But it does require giving up your own will—your independence—and submitting to His rule. You have to love Him back. His yoke is easy, and His burden is light, but you must take on His yoke and His burden if you would live forever.

Jesus said the law and the prophets are summed up in two great commandments. Love God. Love your neighbor as yourself. These are His yoke and His burden.

CHAPTER 15

IS DOUBT SIN?

If what "doubt" means is that we are looking for the truth, trying to find out what is really real, this is certainly not sin. It is honesty. It is truth. Expressing our doubts is authentic. It is real. We have to ask the questions over and over. "Is God real? Is there such as thing as spirit? Is mind different than the physical brain and its electrical functions? Why do we have a conscience? Why do some people not seem to have any conscience? How could God allow this to happen to me? What happens after we die? How do we know?"

Even if we have considered all the evidence and decided that God is real sudden doubts often overwhelm our resolve. It may be that we are washed away by the wave of our naturalist culture in the midst of a biology lecture when all we observe is described with authority as the result of abiogenesis and evolution. It may be a provocative ad on television, a billboard, Facebook, or a store window that carries us away with sexual fantasy. It may be our desire for independence and wealth that takes over our will.

These will happen. It's important to express our doubts to each other. Church should be a safe place to express doubt. It's a form of confession. If you think you are the only one doubting, think again. Jesus was clear. "I have not come to call the righteous, but sinners, to repentance." If you think you are righteous, that you have it all together, Jesus has nothing for you. If you are a doubter and admit it, like his apostle Thomas, He will give you assurance of his reality and truth.

CS Lewis describes faith as the ability to hold on to what you have already decided is true even when assailed by doubt and fear. Do you fear death? Do you worry? Most people do. Most Christians do. Does this startle you? It's based on doubt. This kind of doubt is sin. But don't be dismayed. Everyone sins and falls short of the glory of God. Just go back and try again. If you want God's grace—His forgiveness—He gives it freely.

References

- Jesus and the Eyewitnesses, Richard Bauckham (2006)
- Jesus and the Victory of God, N.T. Wright (1998)
- The Resurrection of the Son of God, N.T. Wright (2003)
- The Historical Reliability of the Gospels, C. Blomberg (1987)
- The Historical Reliability of John's Gospel, C. Blomberg (2002)
- The New Testament Documents: Are They Reliable? F.F. Bruce (2003)
- The Historical Christ and the Jesus of Faith, C. Stephen Evans (1996)
- Warranted Christian Belief, "Two (or More) Kinds of Scripture Scholarship," Alvin Plantinga (2002)
- Is The New Testament History?, Paul Barnett (1986)
- Christian Reflections, CS Lewis (1967)
- The Reason for God, Timothy Keller (2008)
- Science and Evidence for Design in the Universe, by Michael J Behe, William A Dembski, and Stephen C Meyer (2000)
- Darwin's Doubt, Stephen C. Meyer (2014)
- Evolution: Still a Theory in Crisis, Michael Denton (2016)
- The Edge of Evolution: The Search for the Limits of Darwinism, Michael J Behe (2007)
- The God Delusion, Richard Dawkins (2006)
- Breaking the Spell, Religion as a Natural Phenomenon, Daniel Dennett (2006)
- Atheism: A Philosophical Justification, Michael Martin (1989)
- Cosmos, Carl Sagan (1980)
- Contact, Carl Sagan (1986)
- On The Origin of Species, Charles Darwin (1859)
- The Universe Next Door, James W Sire (2009)
- Mere Christianity, CS Lewis (1952)
- The Mystery of Life's Origin: Reassessing Current Theories, Charles B. Thaxton, Walter L. Bradley, Roger L. Olsen (1984)

- Nature's Destiny, How the Laws of Biology Reveal Purpose in the Universe, Michael J. Denton (1998)
- Evolution: A Theory in Crisis, Michael Denton (1986)
- Darwin's Black Box: The Biochemical Challenge to Evolution, Michael J. Behe (1996)
- A Severe Mercy, Sheldon Vanauken (1977)
- Under the Mercy, Sheldon Vanauken (1985)
- The Little Lost Marion and Other Mercies, Sheldon Vanauken (1996)
- For The Glory of God, Rodney Stark (2004)
- On the Genealogy of Morality, Friederich Nietzsche (1887)
- Orthodoxy, GK Chesterton (1908)
- The Practice of the Presence of God, Brother Lawrence (1666)
- "What is Essential In Coming To God," Sermon # 2740, C. H. Spurgeon (www.spurgeongems.org)
- The DaVinci Code, Dan Brown (2003)
- The Gospel Code, Ben Witherington (2004)
- The DaVinci Myth vs. The Gospel Truth, D. James Kennedy (2006)
- Catechism of the Catholic Church. Second Edition, John Paul, Bishop (1994)
- Pilgrim At Tinker Creek, Annie Dillard (1990)
- The Seven Laws of the Learner, Bruce Wilkinson (1992)
- The Purpose Driven Life, Rick Warren (2012)
- Every Young Man's Battle, Stephen Arterburn and Fred Stoeker with Mike Yorkey (2009)
- Every Young Woman's Battle, Shannon Ethridge and Stephen Arterburn (2004)
- As One Devil To Another, Richard Platt (2012)
- The Blessed Life, Robert Morris (2016)
- Miracles, Eric Metaxas (2014)
- Gunning For God: Why the New Atheists are Missing the Target, John C Lennox (2011)
- Seven Days That Divide the World, John C Lennox (2011)

- <u>The Case For Christ</u>, Lee Strobel (1998 & 2016)
- <u>Jesus Through Middle Eastern Eyes</u>, Kenneth E Bailey (2008)
- <u>A Call To The Unconverted</u>, Richard Baxter (1657)
- <u>You Lost Me</u>, David Kinnaman (2011)
- <u>The Holy Spirit And You</u>, Dennis and Rita Bennett (1971)
- <u>The Holy Bible</u>, The New King James Version (1982)
- <u>The Emperor's New Mind</u>, Roger Penrose (1989)
- <u>Fashion, Faith, and Fantasy in the New Physics of the Universe</u>, Roger Penrose (2016)